PLASTICS WASTE

Recovery of Economic Value

JACOB LEIDNER
Ontario Research Foundation
Mississauga, Ontario, Canada

MARCEL DEKKER, INC.   New York and Basel

Library of Congress Cataloging in Publication Data

Leidner, Jacob [date]
  Plastics waste.

  (Plastics engineering; 1)
  Includes bibliographies and index.
  1. Plastic scrap--Recycling. I. Title. II. Series:
Plastics engineering (New York, N.Y.); 1.
TD798.L44   668.4'192   81-2752
ISBN 0-8247-1381-8   AACR2

COPYRIGHT © 1981 by MARCEL DEKKER, INC.   ALL RIGHTS RESERVED

Neither this book nor any part may be reproduced or transmitted in any form or by any means, electronic or mechanical, including photocopying, microfilming, and recording, or by any information storage and retrieval system, without permission in writing from the publisher.

MARCEL DEKKER, INC.
270 Madison Avenue, New York, New York 10016

Current printing (last digit):
10 9 8 7 6 5 4 3 2 1

PRINTED IN THE UNITED STATES OF AMERICA

PREFACE

More than a hundred years have passed since John Wesley Hyatt developed the first American plastic. The growth of the plastics industry, initially very slow, accelerated considerably after World War II. Today the annual production of all types of plastics in the United States is about 30 billion pounds. This phenomenal growth was caused by the desirable properties of plastics and their adaptability to low-cost manufacturing techniques. Although plastics are being used today in practically all areas of consumer products, including construction, transportation, and agriculture, a considerable portion of manufactured resin is still being converted into products having short or intermediate lifetimes. Environmental groups are voicing serious concern about the possible damaging impact of plastics on the environment. One of the most recently publicized examples of the results of such concern is the ban on the use of plastic dishes for school lunches in Tokyo, where pressure from housewives and consumer groups forced the schools to switch to stainless steel.

To a large extent, the attitude of environmentalists toward the industry as a whole is unjustified and is a result of the failure of the plastics industry to develop proper communication with the general public. In the case of the Tokyo schools, the change from disposable to nondisposable utensils does not necessarily mean the reduction of environmental impact: solid waste is only substituted by liquid waste.

The environmental attack on the plastics industry is not justified because of the small contribution of plastics solid wastes (about 2% of solid wastes in North America). It should also be remembered that the large proportion of waste generated by the plastics industry is recycled directly in plant. This process is so widely used and accepted that it is considered to be a normal manufacturing procedure and is rarely labeled "waste recycling." Unfortunately, the easy acceptance of this type of recycling has resulted in a lack of interest in the process by members of the scientific community; thus, understanding of the theoretical basis is rather poor and no generalized theory relating various processing and material parameters to performance of the finished product is available.

Not all plastics waste can, however, be easily recycled. At present a considerable portion of plastics waste is still being disposed of without the recovery of value. However, rising material and energy costs, government regulations, and environmental awareness of the consumer exert a pressure on the industry to change the situation. As new separation and processing technology becomes available, the recovery of value from plastics waste by reprocessing, recovery of energy, and recovery of chemicals will become more and more widespread. The development of new technology will not happen overnight, nor will it be easy or painless. Some of the processes described in this book have never been commercialized and some have been used only temporarily. Some companies mentioned ceased operation after encountering serious technical and marketing problems. On the other hand, there are systems that are operational today and have good potential. Those companies and individuals that persevere through the difficult initial period will create a new and potentially large and important industry.

My objective was to compile material on the various aspects of plastics recycling and, by making it available to the reader, to stimulate new work in the area. If this book becomes obsolete in a very short time, this objective will have been accomplished.

Preface

I would like to express my gratitude to the management of the Ontario Research Foundation, and especially to E. C. Brown, Director of the Department of Materials Chemistry, for making available the funds necessary for the preparation of this work. Special thanks are also due to W. D. Brown of ORF's Information Services for researching the literature and improving the author's style. The cooperation of numerous companies in making technical material available is greatly appreciated.

J. Leidner

CONTENTS

PREFACE		iii
1 INTRODUCTION TO PLASTICS		1
I.	A Brief History of Plastics	1
II.	Basic Concepts of Polymer Science	3
III.	The Common Polymers	6
IV.	From Polymer to Plastic: Plastics Additives	18
V.	From Plastic to Product: Processing of Plastics	27
VI.	Markets for Plastics	40
VII.	Structure of the Plastics Industry	56
VIII.	Prices and Supplies	57
	References	63
2 SOURCES OF PLASTICS WASTE		64
I.	Definitions Regarding Plastics Waste	64
II.	Plastics Cycle	65
III.	Generation of Industrial Plastics Waste	67
IV.	Plastics in Solid Waste	73
V.	Future of Waste Disposal	80
	References	83
3 SEPARATION		85
I.	Separation of Components of Municipal Refuse	85
II.	Separation Processes Specific to Plastics	97
	References	117
4 PRIMARY RECYCLING		119
I.	Degradation of Thermoplastics Due to Repetitive Processing	119
II.	Industrial Practice	134
	References	154

5	SECONDARY RECYCLING	156
	I. Approaches to Secondary Recycling	156
	II. Secondary Recycling by Mechanical Reworking of Plastics Waste	158
	III. Chemical Modification of Mixed Plastics Waste	186
	IV. Secondary Recycling by Coextrusion and Co-injection Molding	206
	V. The Use of Waste Plastics as Fillers	209
	References	215
6	TERTIARY RECYCLING: CHEMICALS FROM PLASTICS WASTE	218
	I. Pyrolysis	218
	II. Chemical Decomposition of Plastics Waste	264
	References	275
7	QUATERNARY RECYCLING: ENERGY FROM PLASTICS WASTE	278
	I. Introduction	278
	II. The Incinerator	279
	III. Examples of Energy Recovery from Municipal Refuse	284
	IV. The Effect of Plastics in Municipal Refuse on the Incineration Process	288
	V. Incineration of Predominantly Plastics Waste	297
	References	305
8	DISPOSAL OF WASTE PLASTICS WITHOUT THE RECOVERY OF VALUE	306
	I. Incineration without the Recovery of Energy	306
	II. Landfill	306
	III. Plastics in Landfill	309
	References	313
	INDEX	315

Chapter 1

INTRODUCTION TO PLASTICS

I. A BRIEF HISTORY OF PLASTICS

Plastics probably have one of the most spectacular growth histories among engineering materials. Within a relatively short time of their inception, these versatile materials began to displace metals and wood in a variety of applications.

In 1900 the only plastic materials available were gutta-percha, shellac, celluloid, and ebonite. A few years later another plastic material became commercialized--casein plastics obtained by reacting milk casein with formaldehyde. In the late 1800s reactions of phenols with aldehydes were of some academic interest. Research in this area led to the development of phenol aldehyde plastics. In a very short time these plastics were well established in a number of applications. In the 1940s the annual production of phenolic resins reached around 175,000 tons. The commercial success of phenolic resins stimulated further research and development effort in the area of plastics. Development of urea-formaldehyde resins took place around 1930. I. G. Farben in Germany produced polystyrene, and Dow Chemical started development work leading to the commercialization of this product. The first patent for the polymerization of polyvinyl chloride (PVC) was taken in 1912, but it took another 15 years to introduce the commercial product. In 1931 polyethylene was accidentally produced by ICI. Eight years later the first polyethylene plant came on stream.

Polymethyl methacrylate was developed before the World War II and was used during the war as aircraft glazing. The war resulted in stepped-up research leading to the development of substitutes for hard-to-obtain natural materials. A crash research and development program was under way in the United States to develop synthetic rubbers. The expertise gained from that work was later important in the development of other polymeric materials. Nylon was developed by a Du Pont team led by W. H. Carothers. In the 1930s nylon was used only to manufacture fibers; the first nylon molding compounds were introduced in the early 1940s. Stepped-up research and development and the new markets created by the war effort resulted in a rapid increase in plastics production. At the end of the war plastics already had a well-established position in the marketplace. After the war the industry directed its efforts toward improving the properties of plastics, and new, more specialized grades soon became available. In the 1950s high-density polyethylenes prepared by the Ziegler and Phillips processes became commercialized, and, shortly after, polypropylene was discovered. A number of new, so-called "engineering resins," such as acetals, polycarbonates, and polysulfones were commercialized [1]. At the same time was witnessed a growth of the plastic additives industry. A large number of specialized additives such as UV (ultraviolet) stabilizers, antioxidants, antistatic agents, pigments, and fillers became available. The development of glass fibers, and a scientific effort leading to the understanding of the theory of reinforcement, resulted in a rapid growth of the reinforced plastics industry. It is expected that future growth of the plastics industry will be directed mainly toward the development of new, more efficient conversion processes, and new, more specialized compounds based on existing polymers.

One can hardly discuss the history of plastics without mentioning the raw materials industry. Early plastics were manufactured from coal and natural products such as milk, cellulose, and molasses. Starting with World War II, a rapid change from the use of natural materials to the use of petroleum products occurred. Today plastics are made almost exclusively from petroleum. About 5% of the petroleum we extract every year goes to petrochemical use. About 1.5% of the total oil and natural gas consumed in the United States each year is used to produce the petrochemical feedstocks for the plastics industry [2]. The oil crisis of 1974 showed how vulnerable the plastics industry is to the disruptions of the oil supply. This realization coupled with the growing concern about the environmental impact of plastics resulted in a number of studies on the disposal of plastics and the utilization of plastic waste. Because of the projected consumption of plastics and the increasing cost of petroleum, the work on recovering value from plastic waste will continue and eventually create a new industry.

II. BASIC CONCEPTS OF POLYMER SCIENCE

A polymer is defined as a large molecule built by the repetition of small, simple chemical units [1]. Table 1.1 shows repeating units of some common polymers. Polymers consisting of the repeating units of one kind are called homopolymers. If various groups are present in the same molecular chain, the polymer is called copolymer. The length of the molecular chain can vary, but materials having a molecular weight less than 10,000 are rarely of technological interest. Each molecular chain may consist of 1,000 to 10,000 or more repeating units. The molecular weight of a polymer is defined by the average molecular weight and the molecular weight distribution (Fig. 1.1).

Out of two examples given in Fig. 1.1, polymer B has a higher average molecular weight and a narrower molecular weight distribution. Average molecular weight can be measured as a number-average or weight-average molecular weight. The ratio of these two is called the polydispersity index. Average molecular weight and the molecular weight distribution are very important since they

TABLE 1.1 Repeating Units of Some Polymers

Polymer	Repeating unit
Polyethylene	$-CH_2-$
Polyvinyl chloride	$-CH_2-CHCl-$
Polypropylene	$-CH_2-CH(CH_3)-$
Acetal	$-CH_2-O-$
Nylon-6/6	$-(CH_2)_4CONH(CH_2)_6NHOC-$
Polystyrene	$-CH_2-CH(C_6H_5)-$

affect processing characteristics and some of the performance characteristics (for example, toughness, strength, solubility, and so on) of the polymer.

The polymers can exist as single molecules or as molecular networks. Single molecules can either be linear or branched, as shown

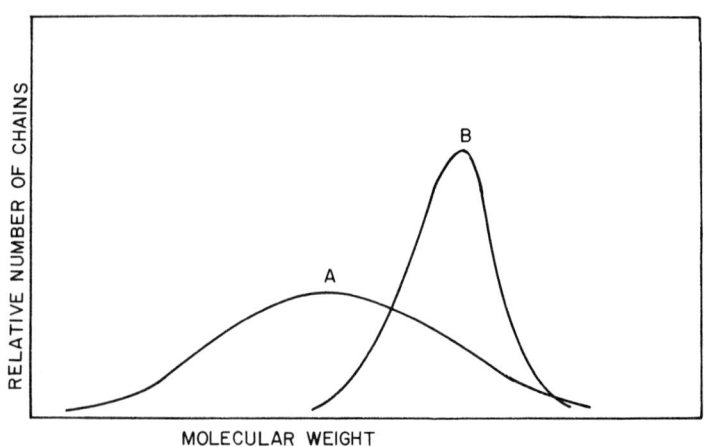

FIG. 1.1 Molecular weight distribution.

Introduction to Plastics 5

in Fig. 1.2. Polymers in the solid state can also be completely amorphous or can show some degree of crystallinity.

An amorphous polymer is often compared to a bucket of worms. The intensity of the movement (segmental Brownian motion) increases with the temperature. Below a certain temperature, known as glass transition temperature, the polymer segments do not have adequate energy to move past one another. The polymer becomes rigid and brittle. Crystalline polymers contain small crystallites which act as cross-links, restraining the movement of the molecular chains and thus contributing to the strength and stiffness of the material.

Plastics made up of single-chain polymers are called thermoplastics. These materials melt on heating and solidify when cooled, allowing them to be processed in the molten state. A molecular network can be achieved by the introduction of cross-links between the molecular chains (Fig. 1.3). If the polymer is only slightly cross-linked, the chains cannot slide past one another, but some flexibility is still retained in sections of the chain remote from the cross-links. At the appropriate

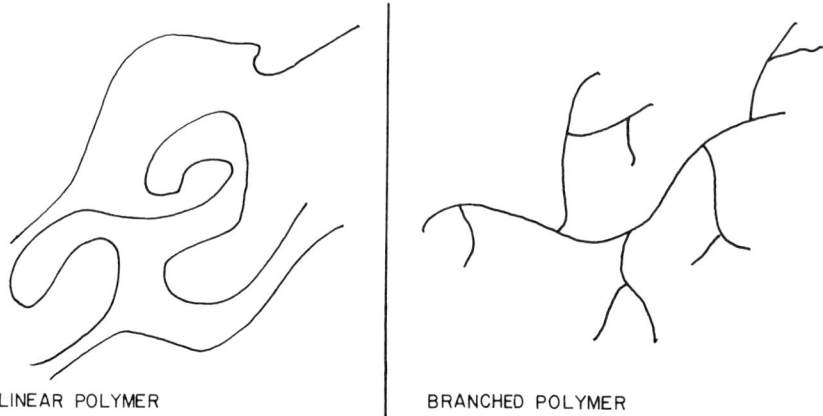

LINEAR POLYMER   BRANCHED POLYMER

FIG. 1.2 Linear and branched polymers.

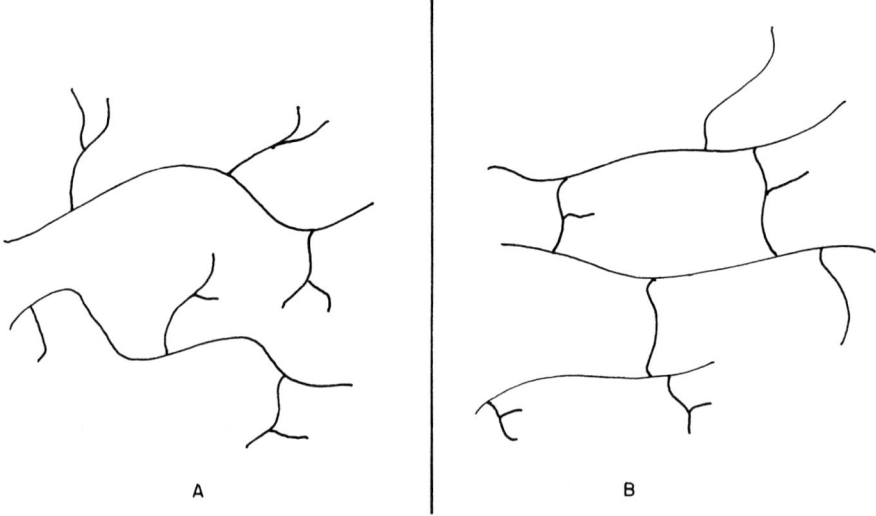

FIG. 1.3 (a) Single-chain polymer and (b) molecular network.

temperature the polymer may be rigid or rubbery. As the degree of cross-linking increases, so does the glass transition temperature (T_g). Eventually the T_g exceeds the decomposition temperature of the polymer. Such a polymer can exist only in a glassy state. Plastics based on polymers containing irreversible cross-links are called thermosets. Once the thermoset has been processed into the final product it cannot be melted and reprocessed.

III. THE COMMON POLYMERS

A. Thermoplastics

Total U.S. production of thermoplastics passed 12 million tons in 1978. Out of that total production over 80% were "commodity resins." The 1977 production breakdown of commodity resins is shown in Table 1.2. Other general-purpose thermoplastic resins such as acrylonitrile-butadiene-styrene (ABS), vinyl acetate, and

Introduction to Plastics 7

TABLE 1.2 Production of Commodity Resin

Resin	1978 Production (1000 tons)	Percent of the market
Low-density polyethylene	3186	25
Polyvinyl chloride	2617	21
High-density polyethylene	1852	15
Polystyrene	1373	11
Polypropylene	1341	11

Based on data from Ref. 3.

acrylonitrile-styrene captured about 11%, and the engineering thermoplastics such as polycarbonates, acetals, and nylons captured about 6% of the total thermoplastics market [3].

1. POLYETHYLENE

Figure 1.4 shows the past and projected growth of low- and high-density polyethylene (LDPE and HDPE). Polyethylenes are manufactured in densities of about 0.91 to 0.96. The density depends on the degree of branching; less branching results in a better molecular "packing" and high crystallinity. High-density polyethylene is more rigid, stronger, and has a higher softening temperature (Fig. 1.5).

There are four methods of producing polyethylene: (1) high-pressure process; (2) Ziegler process; (3) Phillips process; and (4) Standard Oil process.

a. *High-pressure Process*

Polymerization takes place under conditions of high pressure: 1000 to 3000 atm at temperatures of 80 to 300°C. Free radical initiators are used in the reaction. Because the reaction is highly exothermic, a high cooling surface to reactant volume is required. Typically, 10 to 30% of the gas is converted to the

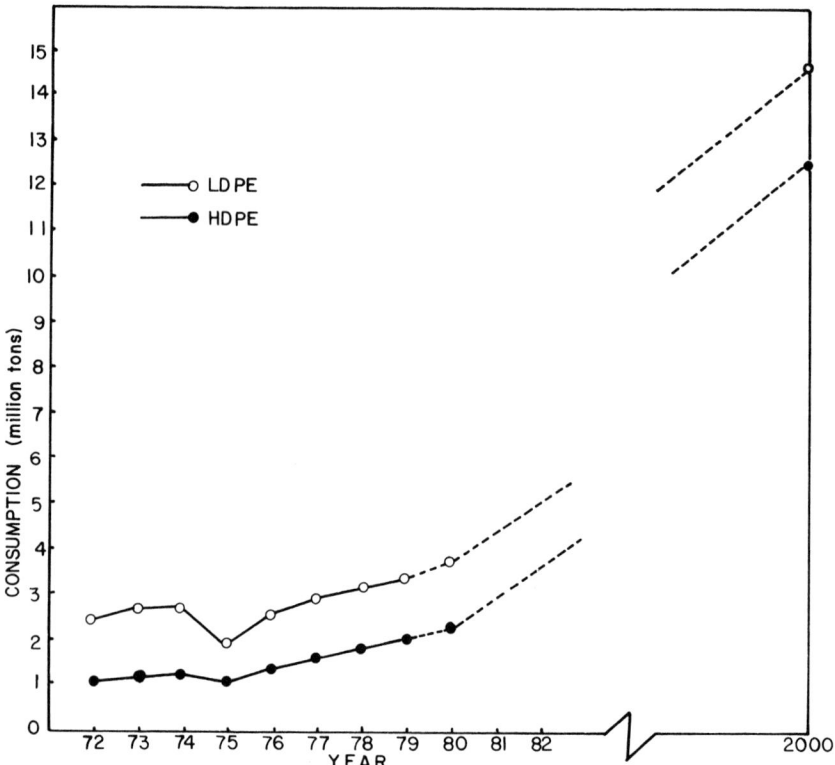

FIG. 1.4 Consumption of LDPE and HDPE. (Data from Refs. 3, 4, and 5.)

polymer. Gas and polymer are then separated and the polymer is pelletized. The high-pressure process is used to manufacture low-density and usually lower-molecular-weight polyethylenes.

b. *Ziegler Process*

This process utilizes coordination catalysts. The reaction takes place at low temperatures (below 100°C) and low pressures. The catalyst complex may be first prepared and then fed into the reaction vessel or can be formed in situ. At the end of the reaction stage the catalyst is separated from the polymer. The Ziegler process is used to manufacture intermediate-density polyethylene (0.945 g/cm^3).

Introduction to Plastics

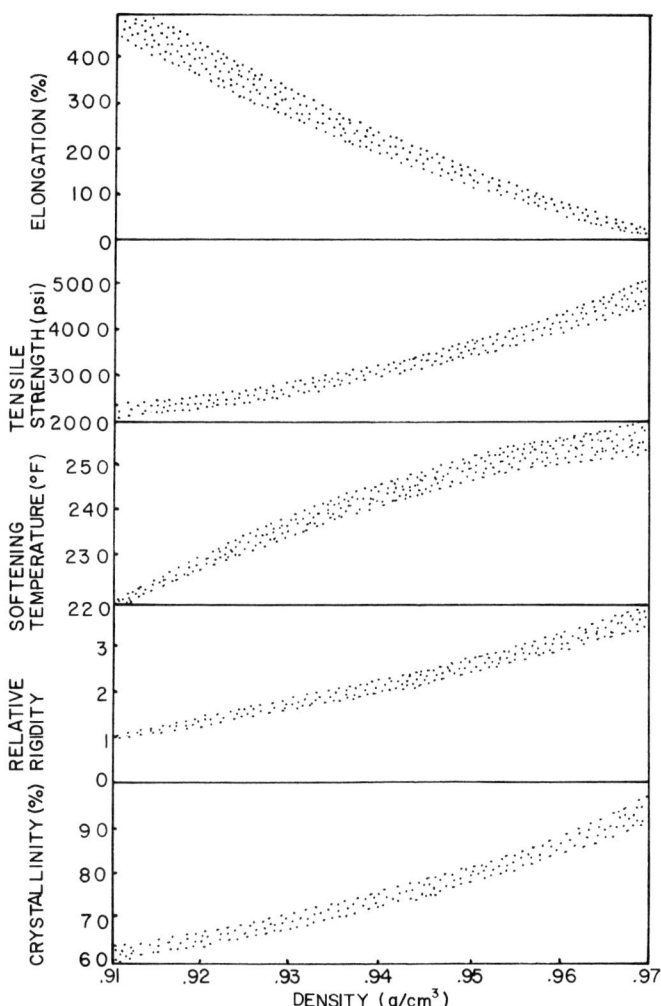

FIG. 1.5 Density of PE versus relative properties.

c. *Phillips Process*

Ethylene dissolved in a liquid hydrocarbon is polymerized by a supported metal oxide catalyst at 130 to 160°C and 200 to 500 psi pressure. Polyethylene made using the Phillips process has a high density (∼0.96). Its molecular weight can be adjusted by changing reaction temperatures.

   d. *Standard Oil (Indiana) Process*

This process is similar to the Phillips process. In both a transition metal oxide is used as catalyst while the reaction takes place at a somewhat higher temperature (230 to 270°C) and lower pressure (40 to 80 atm). High-density polyethylene is prepared by the Standard Oil process.

2. POLYPROPYLENE

Figure 1.6 shows the past and projected growth of polypropylene (PP). Polypropylene has properties similar to those of high-density polyethylene with which it competes for some of the same markets. Some of the important differences are: (1) polypropylene softens at higher temperatures and can be steam sterilized; (2) it is less susceptible to environmental stress cracking but more susceptible to oxidation; (3) it becomes brittle at high temperatures; and (4) it has a lower density (0.9 g/cm^3). Polypropylene is manufactured similarly to polyethylene using Ziegler-type catalysts. Reaction is carried out at 60°C for approximately 8 hr. At the end of the reaction period a mixture of polymer, solvent, monomer, and catalyst is separated and the polymer is blended with antioxidant and pelletized [1].

3. POLYVINYL CHLORIDE

Figure 1.7 shows past and projected growth of polyvinyl chloride (PVC) production. Depending on the type of product desired, vinyl chloride can be polymerized by the suspension process, the bulk process, the solution process, or the emulsion process. Of these the suspension process is the more widely used. In this process vinyl chloride monomer, with the free radical initiator dissolved in it, is dispersed in water. The procedure results in polymer particles of about 100 to 200 μm diameter.

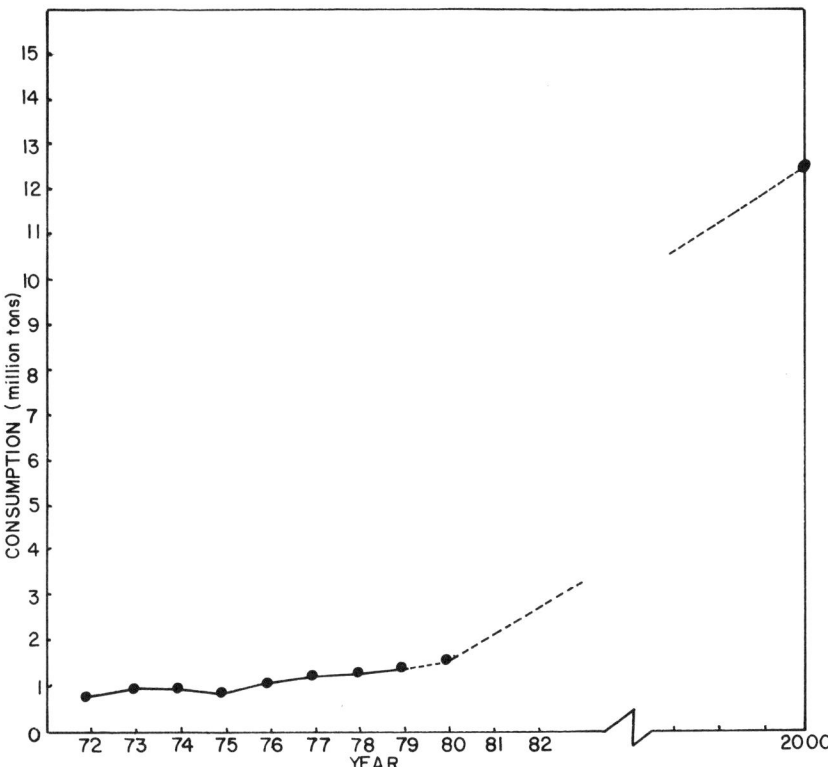

FIG. 1.6 Consumption of PP. (Data from Refs. 3, 4, and 5.)

Bulk polymerization does not require solvents, suspending medium, or surfactants, and it results in a very pure product. The emulsion process utilizes a suspension of monomer in water and a water-soluble initiator. The final product has very small particle size (∿1 μm). The solution process uses a monomer dissolved in a solvent. The polymer, being insoluble in the solvent, precipitates during polymerization.

Polyvinyl chloride by itself has no practical application. The fact, however, that it accepts a wide range of additives makes it one of the most versatile plastics on the market today. It is being used to produce rigid pipe, transparent bottles, and films, as well

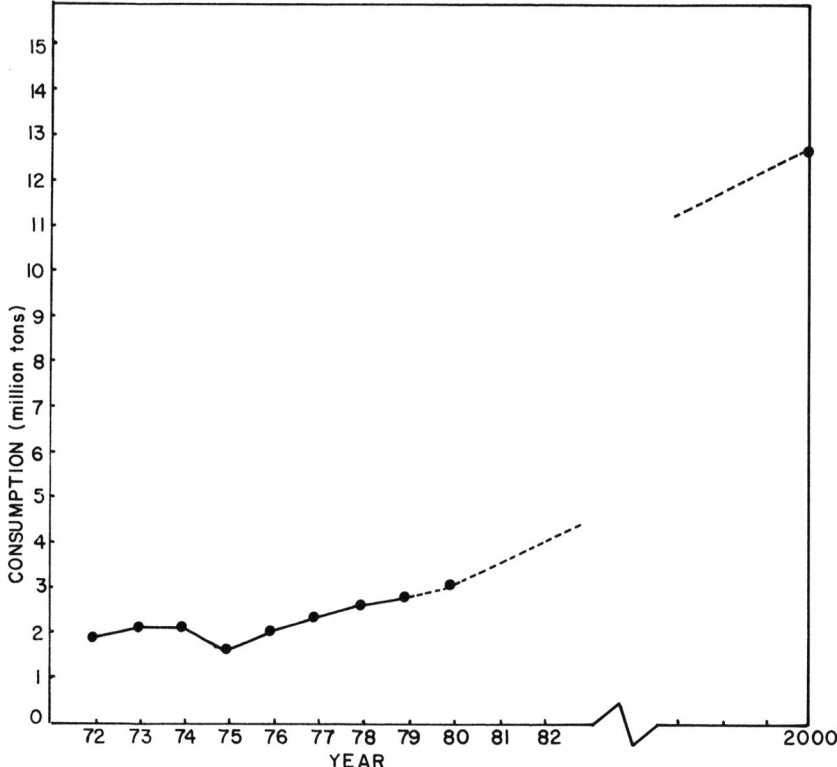

FIG. 1.7 Consumption of PVC. (Data from Refs. 3, 4, and 5.)

as sidings, floor tiles, and artificial leather. The properties of the PVC plastic can be modified by the addition of various stabilizers, pigments, fillers, and plasticizers.

4. *POLYSTYRENE*

Figure 1.8 shows the past and projected growth of polystyrene (PS) sales. Polystyrene is a hard, transparent, and brittle thermoplastic. It is being widely used because it is inexpensive, has low moisture absorption, is easily colored, and is easily moldable.

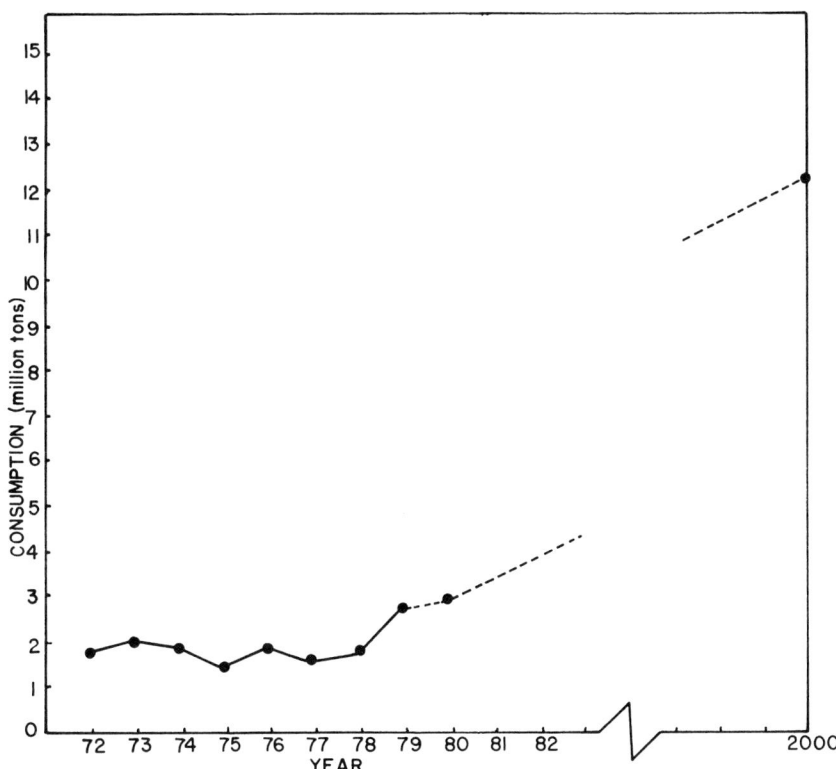

FIG. 1.8 Consumption of PS. (Data from Refs. 3, 4, and 5.)

Six general grades of polystyrene are available:

General-purpose Grades. This category is self-explanatory.

High-molecular-weight Grades. These grades are used when improved impact strength without the loss of optical properties is required.

Heat-resistant Grades. By reducing the amount of the residual monomer, the softening temperature of polystyrene can be increased over the softening temperature of the general-purpose grades.

High-flow Grades. The addition of internal lubricants, a lower average molecular weight and an appropriate molecular weight distribution result in improved flow characteristics.

High-impact Grades. The addition of rubbers to polystyrene results in improved impact absorbing characteristics.

Expandable Grades. Polystyrene granules impregnated with blowing agent (for example, pentane) are being used to make insulating board and a wide range of polystyrene foam products.

Polystyrene, like PVC, is manufactured by suspension, bulk, and solution polymerization techniques.

5. OTHER GENERAL-PURPOSE THERMOPLASTICS

These include polyvinyl acetate (used for emulsion paints, adhesives, paper and textile treatment, and nonwoven adhesives), polyvinyl alcohol, polyvinyl butyral, polyvinylidene formal. The chemistry and uses of these resins are described elsewhere.

6. ENGINEERING RESINS

These materials include polyacetals, polycarbonates, nylons, polyphenylene oxide, polyphenylene sulfide, polysulfone, polyimide, and the thermoplastic polyesters. They are more expensive than the commodity resins and are characterized by excellent mechanical properties, often further improved by the addition of fillers or reinforcing fibers. They are used in such diverse applications as mechanical devices, electrical appliances, plumbing fixtures, and protective helmets. Estimated total 1978 production of engineering resins was 2.85 million tons [3].

7. SPECIALTY THERMOPLASTICS

This group includes acrylics, cellulosics, polyethylene terephthalate (PET, used for beverage bottles, as distinguished from polybutylene terephthalate, PBT, which is an engineering resin), polybutylene, and polyvinyl fluoride. Total sales and use for

Introduction to Plastics   15

all specialty thermoplastics amounted to approximately 700,000 tons in 1978. The use of PET beverage bottles is just beginning. With widespread use predicted for the near future, the influence of PET on specialty thermoplastics statistics could be very great [5].

B. Thermosetting Resins

1. *PHENOLICS*

Phenolics are widely used as molding compounds, adhesives, impregnants, and in many other applications. Since they are among the earliest resins developed, their recent growth has been less spectacular than that of some other plastics. Since no new high-volume uses have been developed for phenolic resins, the growth comes from the expansion of the well-established ones. The 1978 production of phenolics was 658,000 tons [3]. Phenolics are produced by condensation of phenols with formaldehyde.

Two types of phenolic resin are manufactured, the two-stage type and the single-stage type.

   a. *Two-stage Resin (Novolac)*

This resin is obtained by a reaction of phenol and formaldehyde when the amount of formaldehyde is too low to cause complete cross-linking. The resulting produce requires the addition of hexamethylenetetramine (hexa) to complete the cross-linking reaction.

   b. *Single-stage Resin (Resol)*

This resin is obtained by a reaction of phenol and formaldehyde if the amount of formaldehyde is high enough to cause eventual cross-linking. The reaction is stopped before the final cross-linking occurs and is completed later during the molding operation.

Phenolic molding compounds contain a mixture of resin, fillers, lubricants, colorants, and modifiers. These compounds are sold to the molders in the form of a powder or granules.

2. AMINO PLASTICS

The amino plastics are thermosetting resins obtained by reacting of amino-containing compounds (such as melamine or urea) with formaldehyde. These resins are being used in molding compounds, industrial and decorative laminates, and textile-treating resins, and for manufacturing protective coatings and thermal insulating foams [6]. Urea and melamine formaldehyde resins are available in the form of molding compounds containing fillers and modifiers. The 1978 production of these resins reached 575,000 tons [3].

3. UNSATURATED POLYESTER

Unsaturated polyesters are obtained from the condensation reaction of glycols and dibasic acids. The dibasic acid portion of the molecular chain consists of one or more saturated acids and/or unsaturated acid. The unsaturated acid is usually fumaric or maleic. The saturated acid may be isophthalic acid, adipic acid, phthalic anhydride, and so on. The glycols most often used are propylene glycol, dipropylene glycol, ethylene glycol, and diethylene glycol. Unsaturated polyesters are normally solids at room temperature, and to facilitate handling they are dissolved in reactive monomers such as styrene, vinyl toluene, methyl methacrylate, or diallyl phthalate. The addition of monomer causes the blend to become rather unstable, and it has to be inhibited against gelation. Organic peroxides are generally used as the catalysts for initiating hardening (cross-linking) of the compounds.

Unsaturated polyesters are used for appliances, business equipment, consumer goods, transportation, marine applications, corrosion-resistant products, and so on [6].

Total 1978 production of unsaturated polyesters was 524,000 tons [3].

4. EPOXY

Epoxy resins can contain two types of reactive site, the epoxide group and the hydroxyl group. Although epoxies may have a variety of chemical structures, the most widely used are the diglycidyl ethers of bisphenol A (DGEBA). Another important class of epoxy resins (used in flame-retardant applications) is the diglycidyl ether of tetrabromobisphenol A. Except for the presence of bromine, their structures and chemistry are similar.

Epoxies can be cross-linked by a variety of cross-linking agents or hardeners such as modified and unmodified aliphatic and aromatic polyamines, alicyclic polyamines, polyamides, and acid anhydrites. The choice of hardener influences the final properties of the cured compound [6]. Epoxies are used in such applications as coatings, electrical castings, printed circuit laminates, polymer grout, and adhesives. The total usage of epoxy resins reached 141,000 tons in 1978 [3].

5. POLYURETHANES

Polyurethanes are obtained by the reaction of an isocyanate and a polyol. The changes in chemical formulation result in wide variations in physical properties.

Polyurethane products can be divided into three groups; (1) flexible foam, (2) rigid foam, and (3) elastomers. Flexible foam can be produced as a very soft, flexible material replacing latex rubber in home or auto seating, or as a more rigid foam for auto interior padding. Rigid, low-density polyurethane foams have excellent thermal insulating properties while higher-density products find applications as structural parts and decorative moldings. Urethane elastomers are available in a broad range of hardnesses and as both thermosets and themoplastics. Elastomer-type polyurethanes find applications as coatings, adhesives, and sealants [6].

The largest portion of polyurethane is used in the form of foam. In 1978 the production of polyurethane foams reached 840,000 tons [3].

IV. FROM POLYMER TO PLASTIC: PLASTICS ADDITIVES

Most polymers are of little practical value by themselves. They are susceptible to heat and UV degradation, and their physical properties are generally inadequate for specific applications. Parallel to the growth of the plastics industry one can witness the spectacular growth of the plastics additives industry. It is the additives that transfer a polymer (a chemical compound) into an engineering material, a plastic. Growth of the plastics industry means an increasing market for additives; and the development of new additives contributes to the processing and performance values of plastics, thus stimulating the growth of the plastics industry. Plastics additives are very diverse products, each group having its own separate scientific and technological background. Some groups of additives are briefly described.

A. Property Modifiers and Processing Aids

This group of additives includes impact modifiers (or tougheners), flow modifiers, lubricants, and release agents.

1. IMPACT MODIFIERS

These materials, usually elastomers, improve impact and crack resistance when compounded with the base resin. Solid elastomers are used with thermoplastics while liquid reactive rubbers are used with thermosets.

2. FLOW MODIFIERS

These materials, when added in small amounts to the base resin, reduce the melt viscosity resulting in faster production rates.

Introduction to Plastics

Typical examples of flow modifiers are polybutylene resins (for use in polyethylene or polypropylene), an ethylene-acrylic acid copolymer (for use in nylon), and low-molecular-weight polyamide (for use in nylon and thermoplastic polyester).

3. LUBRICANTS AND RELEASE AGENTS

These materials have a tendency to migrate to the surface of the plastic melt during the processing, reducing friction and providing a release layer between the plastic and the mold. The most widely used release agents are metallic stearates, followed by waxes, fatty acid amides, and fatty acid esters.

B. Colorants

Color is extremely important in broadening consumer acceptance of plastics; both organic and inorganic types are used. The following forms of colorants are used by the plastics processor:

Powder or Granules. The product contains mainly colorant with little or no binder.

Pellet Concentrates. The colorant is predispersed in the polymer. The processor dilutes this concentrate during the processing. The advantage for the processor in using pellet concentrates is the ease of dispersion and lack of dust.

Liquid Concentrates. Colorant is suspended or dissolved in the liquid and is continuously metered into the polymer melt during the processing. A special color dispenser is required for liquid concentrates.

Table 1.3 shows the amounts of the colorants consumed by major plastics groups in 1978.

C. Plasticizers

The plastics industry uses plasticizers almost exclusively in PVC. Plasticizer when incorporated in PVC reduces its glass transition

TABLE 1.3 Colorants Consumed by Major Plastics Groups, 1978

Market	Resin containing colorant (1000 tons)	Colorants consumed (tons)
LDPE	1,403	26,980
HDPE	1,141	13,185
PVC	1,910	32,075
PS	1,345	19,880
PP	897	11,030
ABS	365	7,760
Total	7,061	110,910

Reprinted from Ref. 7, courtesy McGraw-Hill.

temperature and converts a hard, rigid polymer into a flexible, soft one. Total PVC consumption of plasticizers in 1978 was 597,000 tons [7]. Plasticizers are also much used in cellulosics, but the market is relatively small.

D. Antioxidants

Most of the polymers are susceptible to oxidation. This susceptibility can be influenced by the polymerization process, processing conditions during the conversion process and the end-use conditions. Two general groups of antioxidants are used: (1) primary or chain terminating antioxidants, which function by transfer of hydrogen to free radicals; (2) secondary or peroxide-decomposing antioxidants, which destroy hydroperoxides, the sources of free radicals in polymers [6]. Table 1.4 shows the amounts of antioxidants used in various plastics.

Introduction to Plastics

TABLE 1.4 Antioxidant Consumption by Plastics, 1978

Plastic	Antioxidant consumption (tons)
ABS	4,830
PE	3,140
PP	4,220
PS	2,410
Others	1,410
Total	16,010

Reprinted from Ref. 7, courtesy McGraw-Hill.

E. UV Stabilizers

Most polymers are sensitive to sunlight. Especially damaging is lower-wavelength radiation (approximately 290 µm). UV degradation of plastics is manifested by yellowing, cracking, and embrittlement. UV stabilizers can protect polymers, (1) by absorbing UV radiation and emitting longer-wavelength (harmless) energy, (2) by absorbing UV radiation and breaking down in the process (these stabilizers are slowly depleted in the process), and, (3) by making polymer opaque to UV radiation (some pigments act that way). Table 1.5 shows the quantities of stabilizers used in plastics during 1978.

F. Flame Retardants

Flame retardants modify the combustibility characteristics of plastics. Properties such as flame spread, ignition temperature, and rate of burning can be controlled to some extent with flame retardants.

TABLE 1.5 Use of UV Stabilizers, 1978

Material	Use of stabilizers (tons)
Acrylics	61
Cellulosics	15
Polycarbonates	107
Polyesters	63
Polyolefins	1472
Polystyrenes	55
Polyvinyl chloride	61
Coatings	52
Others	29
Total	1915

Reprinted from Ref. 7, courtesy McGraw-Hill.

The following classes of flame retardants are being used:

1. Organic additive flame retardants:
    Chlorinated hydrocarbons
    Brominated hydrocarbons
    Organophosphorus compounds

2. Inorganic additive flame retardants:
    Antimony oxide
    Zinc oxide
    Zinc borate
    Molybdenum oxide
    Alumina trihydrate

Introduction to Plastics

3. Organic reactive flame retardants:
   Halogenated polyols
   Brominated aromatics
   Phosphorus polyols

Table 1.6 lists the amounts of various flame retardants used in plastics in 1978.

G. Fillers

Fillers play a double role in plastics: they are inexpensive extenders used to reduce an overall cost of the compound and as property modifiers to improve such properties as flow uniformity, stiffness, and heat distortion temperature. Very often the improvement of one property is at the expense of another.

Following are the main groups of fillers used in plastics:

1. CALCIUM CARBONATE

This filler is available in natural and synthetic forms. Calcium carbonate is an inert, nonreinforcing filler. However, when present in concentrations greater than 10%, it does increase the tensile strength of composites to some extent. Some surface-treated varieties of calcium carbonate fillers also contribute to an increase in the tensile strength of some resins. Calcium carbonate filler is also used with PVC, polyolefins, epoxy, and phenol-formaldehyde resins [8].

2. SILICAS

Silicas occur in many forms: quartz, sand, diatomaceous earth, novaculite, and many others. The properties and uses of silicas are numerous. Sand with phenolic binder is used as shell molds for casting metals. Silica flour is used as an inert filler in both thermoplastics and thermosets. Very fine silica is used as a thixotropic agent.

TABLE 1.6 Flame Retardants Used in Plastics, 1978

Type	1978 Consumption (in thousand tons)
Additives	
Alumina hydrates	90
Antimony oxides	15
Boron compounds	5
Bromine compounds	13
Chlorinated paraffins and cycloaliphatics	17
Phosphate esters	18
Others	7
Total	175
Reactive systems	
Epoxy intermediates	5
Polycarbonate intermediates	2
Polyester intermediates	7
Styrenic intermediates	2
Urethane intermediates	12
Others	3
Total	31

Reprinted from Ref. 7, courtesy McGraw-Hill.

3. *KAOLIN*

Kaolin (a form of clay) is being used both in natural and calcined form. In both forms it has broad applications as a filler in plastics (both thermoplastics and thermosets) and in rubber. It is characterized by good chemical resistance and electrical properties. Surface-treated fine kaolin particles increase strength and modulus of some thermoplastics [6].

4. ALUMINA TRIHYDRATE

Alumina trihydrate is widely used, especially in some thermosets such as polyester and epoxies, because of its flame-retarding properties.

5. OTHER FILLERS

Fillers such as glass spheres, hollow glass spheres, wood flour, cellulose fibers, and ground mica are also used by the plastics industry.

Table 1.7 shows past and projected consumption of some fillers.

TABLE 1.7 Consumption of Fillers in Plastics

Type of filler	Consumption (1,000 metric tons)			
	1975	1980 est.	1990 est.	2000 est.
Calcium carbonate	700	1,500	2,500	9,000
Talc	40	200	900	1,800
Alumina trihydrate	50	200	800	1,600
Asbestos	180	350	800	1,700
Cellulosics	40	90	300	500
Silica	25	80	300	500
Silicates	6	15	50	100
Total	1,041	2,435	6,650	15,200

Reprinted from Ref. 8, courtesy Marcel Dekker.

H. Reinforcements

Reinforcements are used to improve such mechanical properties of plastics as tensile strength, Young's modulus, impact strength, and heat distortion temperature. Reinforcements can be distinguished from fillers by their geometry: fillers have irregular shapes while reinforcements are fibers or flakes. Reinforcements are either continuous or discontinuous. The most widely used reinforcements are glass fibers. They can be used both as continuous reinforcements (with thermosetting resins) or as discontinuous ones (chopped or milled fibers, in both thermosets and thermoplastics). High-aspect-ratio (HAR) mica flakes are a rather recent addition to the family of reinforcements. Table 1.8 shows the 1977 pattern of consumption of reinforced thermoplastics and Fig. 1.9 shows the projected growth of fillers and reinforcements for plastics.

TABLE 1.8 Pattern of Consumption of Reinforced Thermoplastics, 1978

Material	Market (1000 tons)
Nylon	19
Polyester	22
Polyethylene	
Polypropylene	34
Styrenics	14
Other	11
Total	100

Note: Reinforcements are used at the level of 10 to 40%. Reprinted from Ref. 3, courtesy McGraw-Hill.

Introduction to Plastics

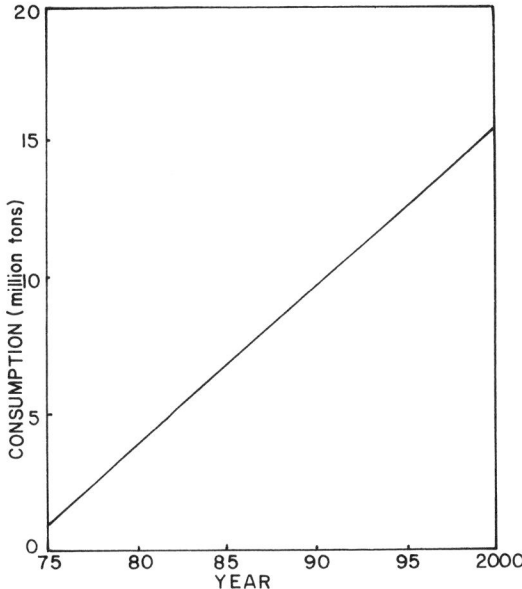

FIG. 1.9 Growth of consumption of fillers and reinforcements for plastics. (Data from Ref. 8.)

V. FROM PLASTIC TO PRODUCT: PROCESSING OF PLASTICS

Plastic products are produced by numerous processes, only the most popular of which are discussed here.

A. Extrusion

Figure 1.10 schematically shows a single-screw extruder. Resin in the form of powder, pellets, flakes, beads, or granulated regrind (often premixed with colorants, stabilizers, and lubricants) is fed to the extruder hopper. From the hopper the resin is picked up by the screw which conveys it to the die. In the barrel the material is heated or cooled and develops the necessary pressure to force the melt through the die. The most common extrusion processes are: (1) compounding and pelletizing, (2) pipe and

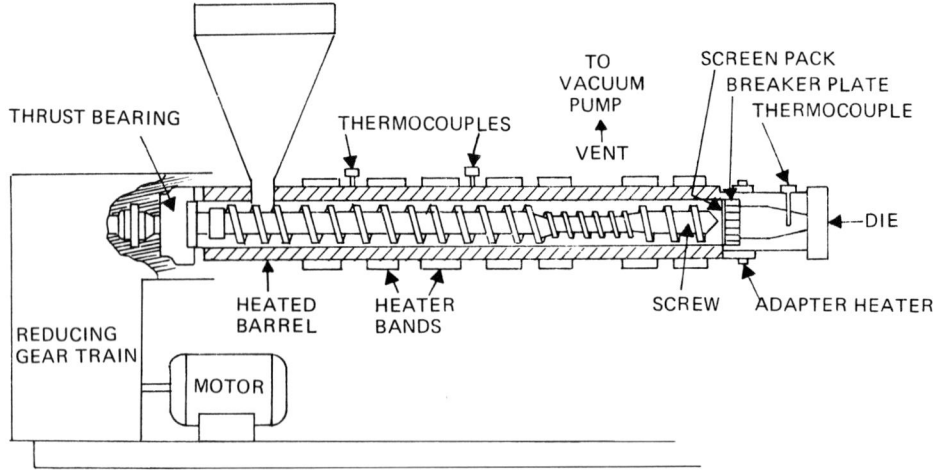

FIG. 1.10 Extruder. (Reprinted from Ref. 15.)

tube extrusion, (3) profile extrusion, (4) wire and cable coating, (5) film extrusion, and (6) extrusion blow molding.

1. COMPOUNDING AND PELLETIZING

In order to prepare resins in the pellet form, the resin manufacturer can charge the polymer, with or without the appropriate additives, directly into the extruder which pumps the molten resin through a die face cutter system. A large number of strands leaving a pelletizing die are cut by blades operating at high speeds against the die plate face. The operation is conducted under the water. The slurry of the pellets and water is conveyed into the drier where the water is removed.

Another method used to produce pellets, strand pelletizing, uses a strand die, water-cooling die, water-cooling tank, water stripper, and cold strand pelletizer. After exiting from the extruder, the strands are cooled in the water tank, dried in the water stripper, and pelletized [6].

2. PIPE AND TUBING

A typical pipe extrusion system incorporates a vacuum cooling tank (vacuum sizer), a water spray or submerged cooling tank, a pipe puller, a pipe cutoff saw, and a dump table. In-line printing equipment is also frequently used.

The hot plastic tube leaving the extruder has poor strength, and will collapse or distort if submerged in water. To prevent this a vacuum sizer is used to create a differential pressure, which inflates the pipe against the sizing sleeve, so that the pipe retains its shape while being cooled. Tubing is manufactured in a similar fashion, although vacuum sizing is not always used [6].

3. PROFILES

Profiles are usually cooled by spray or immersion tanks (solid profiles) and in vacuum tanks (hollow profiles). If slower cooling is required, air cooling can be used. Shape-holding devices are used to assure that specified dimensional tolerances are met [6].

4. SHEET EXTRUSION

The extrusion of a sheet consists of the following steps:

1. A sheet die mounted on the extruder deposits plastic melt in sheet form on a polishing roll assembly (usually a stock of three vertical, highly polished, water-cooled, motor-driven rolls).
2. The rolls gage the sheet to the required thickness, cool it, and apply the desired surface.
3. The sheet is conveyed over rollers for air cooling and passed through motorized rubber draw rolls.
4. The sheet is cut to the desired length [6].

5. FILM EXTRUSION

Films are mainly produced by extrusion casting process and extrusion blown film process. In extrusion casting of film (Fig. 1.11), a flat die produces a thin web of molten film which is deposited onto a cooling drum. After cooling and solidifying, the film typically passes through pull rolls, slitters, winders, and printing equipment.

Figure 1.12 shows the blown film process. From a tubular die the product, in a form of a thin tube, moves upward to a film tower (which contains a collapsing frame, guide rolls, and motor-driven pull rolls). The tube is inflated by air passing through the center of a die. The outside surface is cooled by air from an air ring mounted above the die. After the tube passes through the pull rolls it may be slit to form two separate sheets, each of which is separately wound on the roll. [6]

6. COEXTRUSION

Coextrusion allows the formation of film or sheet laminates, with each layer contributing different properties to the overall

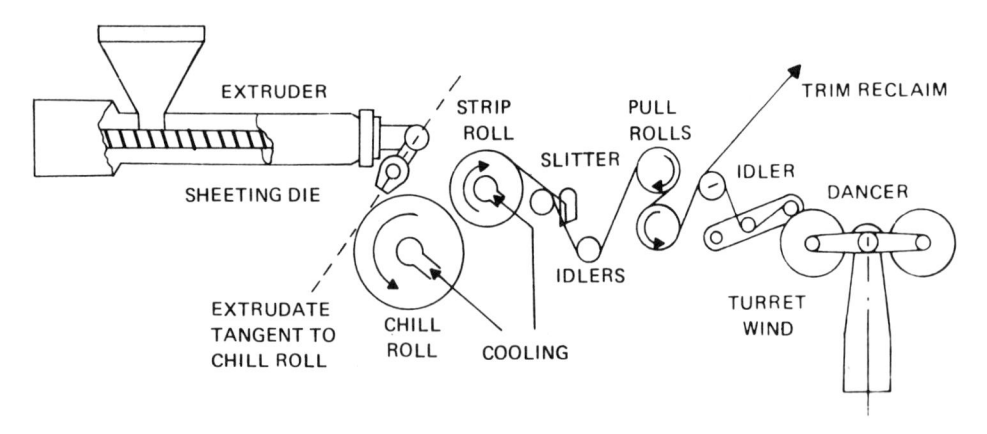

FIG. 1.11 Extrusion casting of plastic film. (Reprinted from Ref. 15.)

Introduction to Plastics

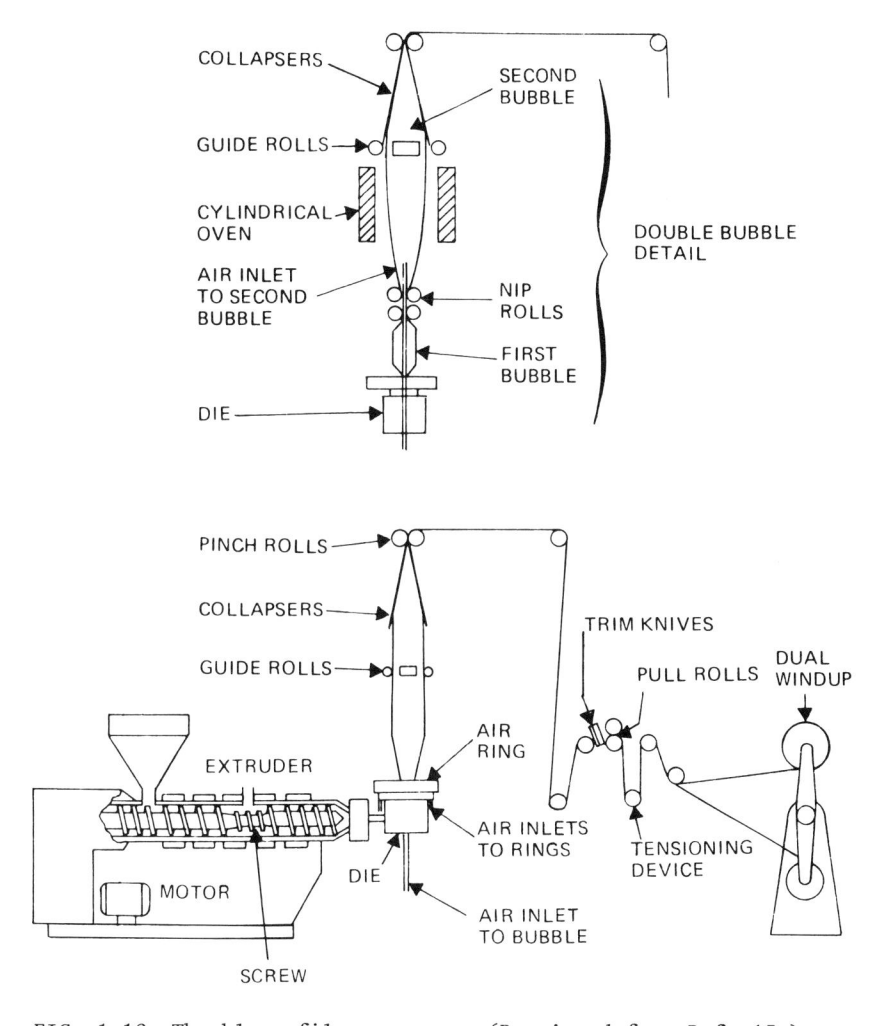

FIG. 1.12 The blown film process. (Reprinted from Ref. 15.)

properties of the product. Different plastics can be combined in a film or sheet structure to achieve, for example, good clarity, tear resistance, and impermeability to water vapor. The combining of layers takes place in a combining adapter with a single-manifold die or at the exit of a multimanifold die [6].

7. WIRE AND CABLE EXTRUSION

A typical wire coating line consists of (1) wire payoff, (2) wire preheater, (3) extruder, (4) die, (5) cooling trough, (6) capstan, (7) accumulator, and (8) wire take-up. A wire is unwound from a reel, preheated to improve adhesion, and then coated with plastic in the extrusion die. Plastic insulation is cooled in a cooling trough. A capstan draws the wire through the operation and an accumulator provides for inventory coated wire. The take-up winds the wire on the reels [6].

B. Injection Molding

Two types of injection-molding systems are in use: plunger and reciprocating screw machines. Reciprocating screw molders are the most widely used, comprising over 75% of all injection machines in operation today.

Figure 1.13 schematically shows a representative injection molding machine. Resin in the form of pellets or sometimes powder is fed from the hopper into the heated barrel. A rotating screw conveys the plastic forward and plasticates it. The screw itself moves back, allowing the accumulation of the melt in the front portion of the barrel. When an adequate amount of the material has accumulated, the screw stops rotating and moves forward, pushing the melt through the sprue bushing, and the runners and gates into the mold cavity. After the melt has cooled and solidified, the mold opens and the part is ejected. Sprues and runners are removed from the parts, reground, and usually reused.

Injection molding can be used for both thermoplastics and thermosets. In the case of thermoplastics, material is melted in the heated barrel and solidifies on cooling in the mold. Thermosets are plasticated in a slightly warmed barrel and solidify in a heated mold. Approximately 50,000 injection molding machines are presently in operation in the United States, processing some 23% of all plastics in the country.

Introduction to Plastics   33

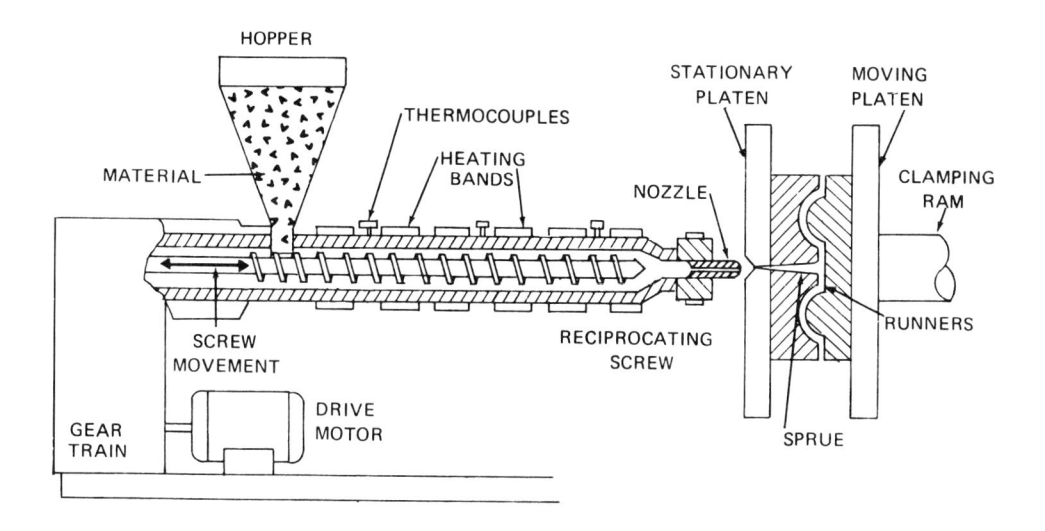

FIG. 1.13 Injection-molding machine. (Reprinted from Ref. 15.)

C. Blow Molding

There are three basic types of blow molding in use today:
(1) extrusion blow molding, (2) injection blow molding, and
(3) stretch blow molding. Figure 1.14 schematically shows extrusion blow molding. The process consists of four steps:
(1) melting and plasticating the resin in the extruder,
(2) extrusion of the tube (parison), (3) closing the mold
around the parison and inflating the parison, (4) opening the
mold, removing the part from the mold, and removing the flash.

   Figure 1.15 shows the basic principles of injection blow
molding. In injection blow molding the parison is injection
molded on a steel core. The parison is then positioned in a
blow mold where the bottle is blown without any scrap. The
distribution of the material is determined by the injection
molding of the parison which makes the whole process more uniform
and more automatic [6].

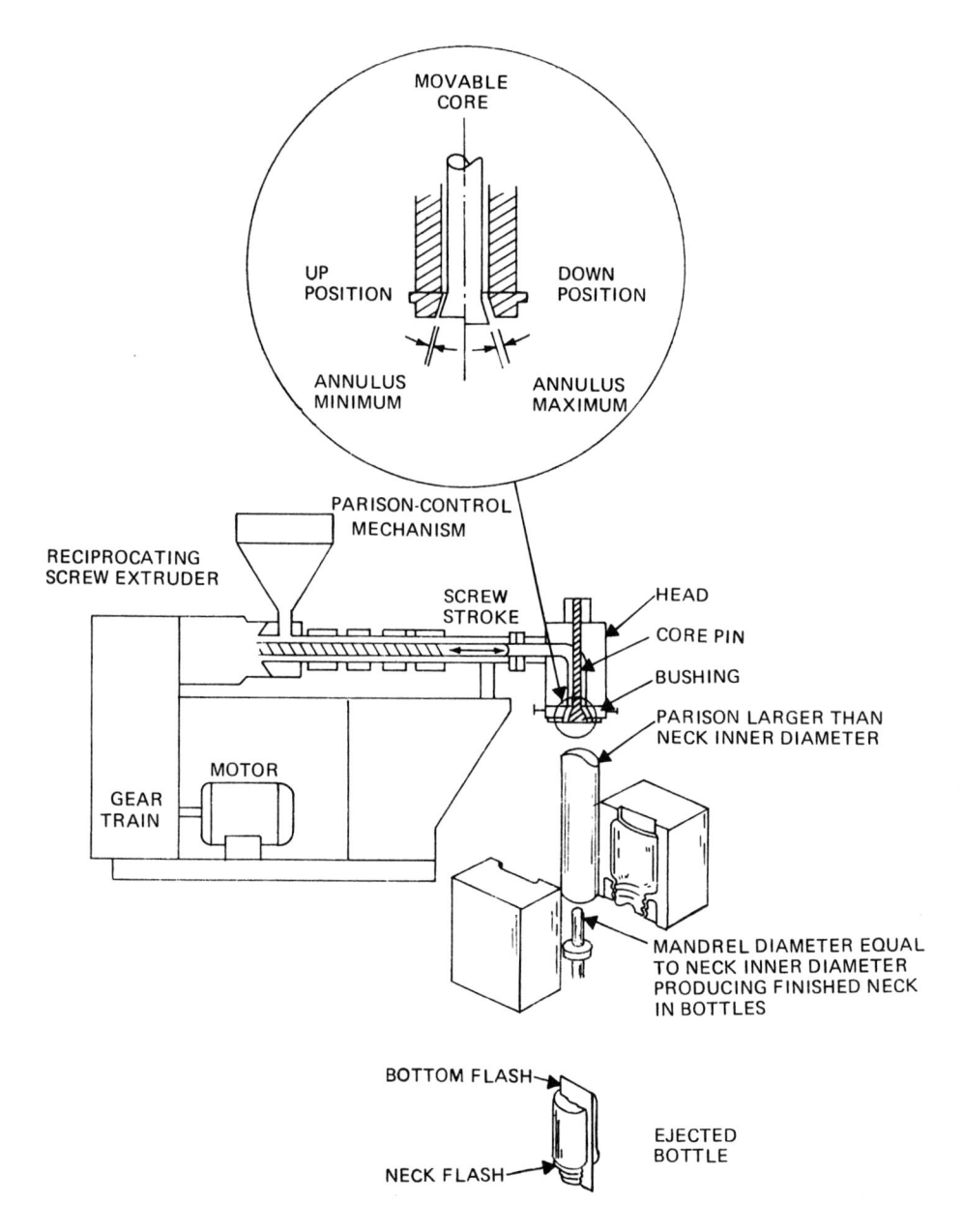

FIG. 1.14 Extrusion blow molding. (Reprinted from Ref. 15.)

Introduction to Plastics

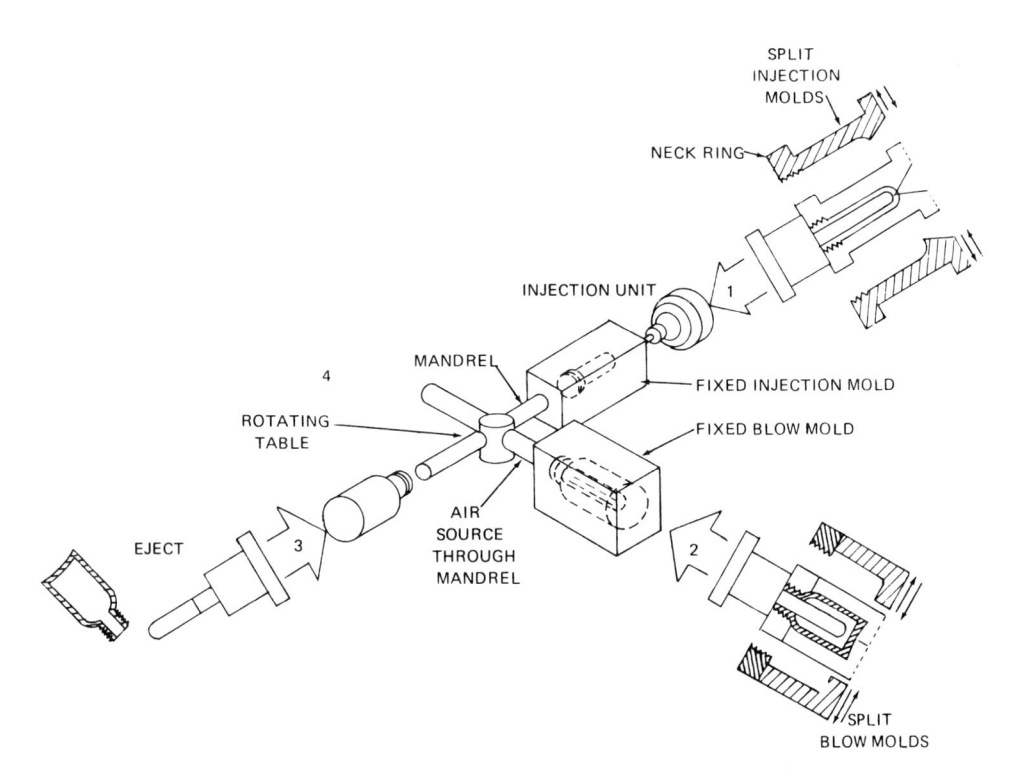

FIG. 1.15 Four-station rotary injection blow molding machine:
(1) complete injection molding of parison; (2) close mold, blow,
and cool; (3) eject; (4) cool mandrel. (Reprinted from Ref. 15.)

Stretch blow molding introduces biaxial orientation to the material (aligns molecules in two planes), resulting in improved strength, better contact clarity, better impact strength, improved gas and water vapor barriers, and reduced creep. In biaxial orientation the bottle is stretched in the radial direction by compressed air, and mechanically along the vertical axis. Stretch blow molding is gaining popularity, especially when used to manufacture polyethylene terephthalate (PET) and polypropylene containers. PET has been proven in carbonated beverage bottles and in other food packaging. Improved clarity of biaxially oriented PP opened up markets for PP containers in toiletry, cosmetic, detergent, and pharmaceutical markets [6].

D. Rotational Molding

Rotational molding is used mainly for the production of hollow objects from thermoplastics and to a lesser extent, from thermosets. The equipment utilized in this process is relatively low-cost and durable. The basic concept of rotational molding is shown in Fig. 1.16.

Plastic powder or liquid is placed in the molds, which are then rotated while being heated. Powder melts and coats the surface of the mold. If reactive liquid resin is used, it solidifies on contact with the hot surface. In either case the mold must be cooled before the part can be removed.

E. Compression and Transfer Molding

Both of these processes are used to mold thermosets. In compression molding (Fig. 1.17) a molding compound is placed in the cavity of the heated mold, which is then closed. The heat and pressure cause the molding compound to flow and fill the cavity. After the material solidifies the mold is opened and the part removed. Transfer molding (Fig. 1.18) differs from compression molding in that the molding compound is preheated and softened in a separate chamber and then pushed into the mold cavity by a plunger.

F. Calendering

Among plastic materials, PVC flexible sheet and film account for the bulk of calendering output. Other materials processed by calendering include rigid, expanded, polished and clear vinyls, and limited amounts of ABS and PE. Approximately 150 calendars were operating in 1979 in the United States.

Introduction to Plastics

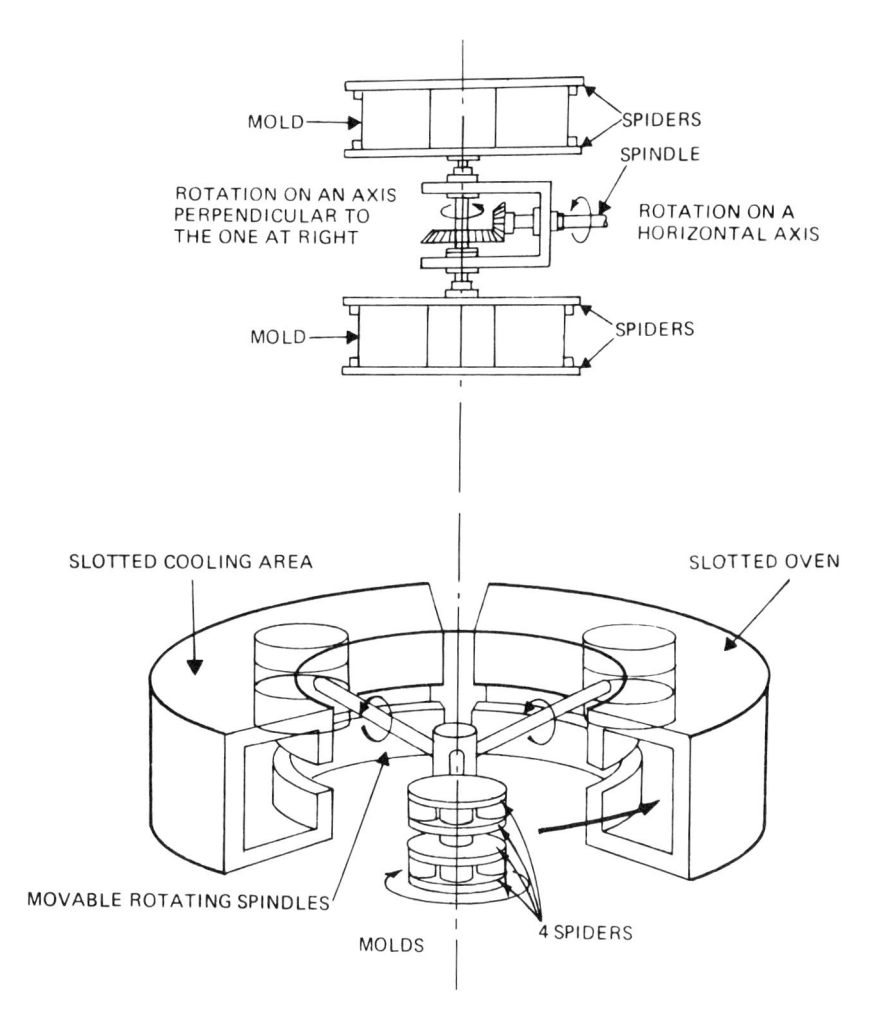

FIG. 1.16 Rotational molding. (Reprinted from Ref. 15.)

Figure 1.19 shows a typical four-roll calendar installation. Calendering is a form of extrusion, the rolls forming rotating die lips. Friction between the molten resin and the roll faces, and the viscosity and elasticity of the melt, force the material to extrude through the gap between the rolls. The other four rolls

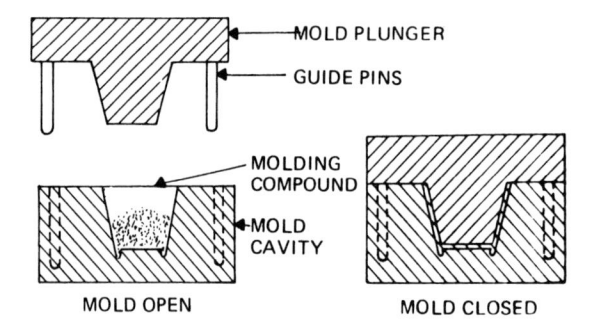

FIG. 1.17 Compression molding. (Reprinted from Ref. 15.)

form three nips: a feed pass, a metering pass, and a sheet forming, gaging, and finishing pass [6].

G. Thermoforming

In the thermoforming process thermoplastic sheet is formed into a product by first being softened by the applications of heat, and then shaped by the application of pressure and by pressing the hot

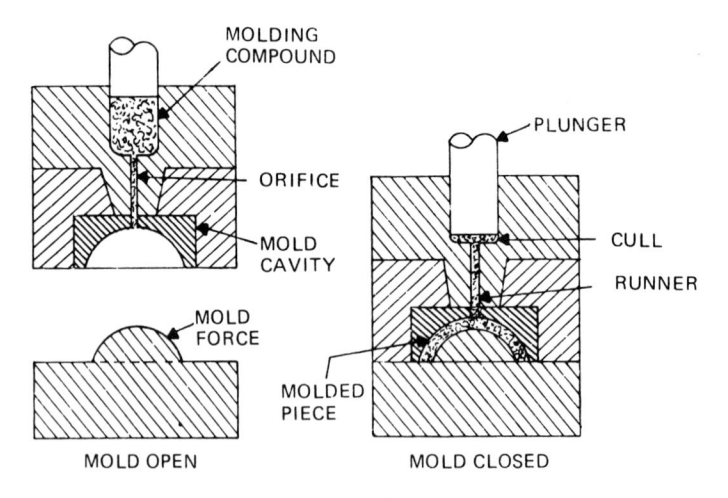

FIG. 1.18 Transfer molding. (Reprinted from Ref. 15.)

Introduction to Plastics 39

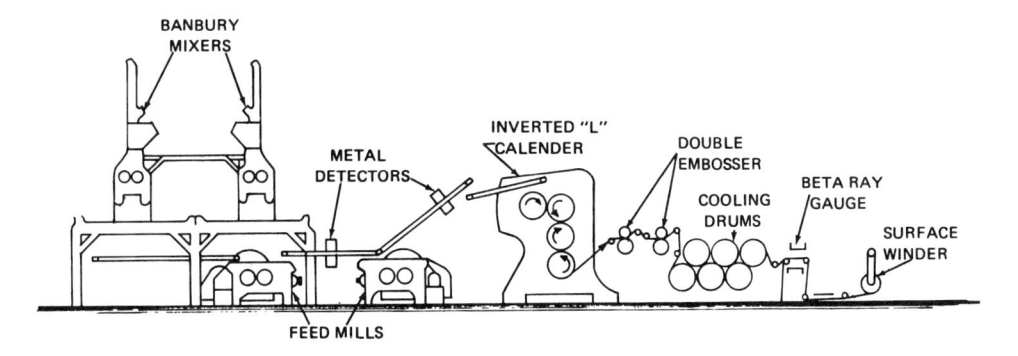

FIG. 1.19 Calendering. (Reprinted from Ref. 15.)

sheet against the cold mold. Forming pressure may be developed by vacuum, compressed air, or by mating matched molds. The thermoforming process consists of four steps: heating, forming, cooling, and trimming. Trimming does not have to be an integral part of thermoforming process, but in only a few instances can it be avoided altogether [2]. Figure 1.20 shows a schematic of the straight vacuum-forming process.

H. Film Lamination

Films can be combined with other types of films or substrates such as aluminum foil or paper to achieve a specific set of cost/performance characteristics. There are a number of methods for laminating films, among which are thermal dry bonding and adhesive laminating.

In thermal dry lamination no adhesives are used. Films such as LDPE, HDPE, and PVC are combined under heat and pressure. However, when it is desired to facilitate bonding, preapplied adhesive coatings can be used. Films laminated by the thermal dry lamination technique are characterized by good clarity but low bond strengths. The process itself is relatively slow due to the slow transfer of heat, and the commercial importance of the process is decreasing.

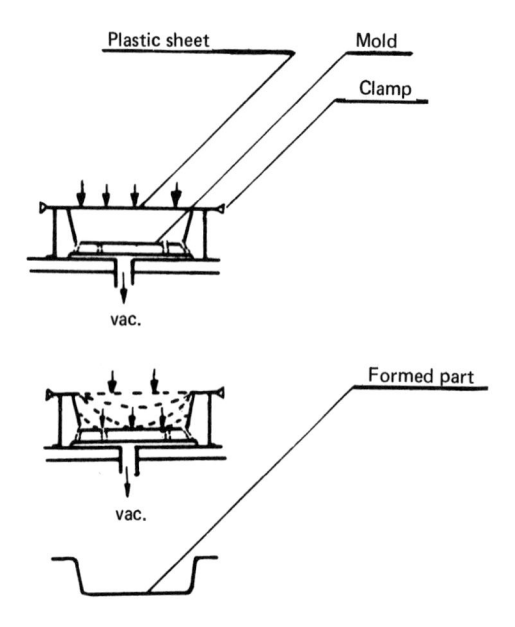

FIG. 1.20 Vacuum forming.

Adhesive dry-bond lamination is becoming dominant because of its simplicity, high productivity of the equipment, and the broad variety of films that can be laminated.

In this process, adhesive in the form of a solution or emulsion is deposited onto the substrate and the solvent is allowed to evaporate. The coated substrate is then combined with another substrate by pressure alone or by pressure and heat [6].

VI. MARKETS FOR PLASTICS

The main markets for plastics are

Building and construction
Packaging
Transportation

Introduction to Plastics 41

Furniture
Electrical and electronic applications
Housewares
Appliances

Of these markets building and construction is the largest single one, consuming in 1978 3.2 million tons of plastics, approximately 20% of the total plastics production. Table 1.9 summarizes growth of markets for plastics between 1974 and 1978.

A. Building and Construction

Table 1.10 shows the 1978 usage of various plastics in building and construction. If the use of plastics in building and construction keeps increasing at the present rate, plastics could become

TABLE 1.9 Use of Plastics by Market, 1974-1978 (in 1000 Metric Tons)

Market	1974[a]	1978[b]	Major types
Building and construction			PVC, phenolic, alkyd, urea, polyester, PE, styrenics, acrylics
Packaging	2671	3564	PE, styrenics, PVC, PP
Transportation	658	1877	PVC, PP, urethane, polyester, ABS
Furniture	493.9	476	PVC, styrenics, urethanes
Electrical and electronic	755.3	782	Phenolic, PE, PVC
Housewares	585	655	PE, styrenics, PVC
Appliances	417.3	399	Styrenics, phenolic, PP, urethane

[a]From Ref. 9.
[b]From Ref. 3.

TABLE 1.10 Plastics in Building, 1978

Application/material	Consumption (1000 tons)
Decorative laminates	
Phenolic	20
Urea and melamine	17
Flooring[a]	
Epoxy (including paving)	8
PVC	147
Urethane foam (rug underlay)	60
Glazing and skylights	
Acrylic	32
Reinforced polyester[b]	15
Polycarbonate	33
Insulation	
Phenolic (binder)	125
Polystyrene foam	57
Urethane foam (rigid)	110
Lighting fixtures	
Acrylic	10
Cellulosics	2
Polycarbonate	5
Polystyrene	12
PVC	7
Panels and siding	
Acrylic	6
Butyrate	2
PVC	61
Reinforced polyester[b]	65
Pipe, fittings, conduit	
ABS	132
Epoxy (coatings)	4
HDPE	229
LDPE	11
PP	9
PS	6
PVC	985
Reinforced polyester[b]	97
Profile extrusions[c]	
PVC (including foam)	70
PE	3

TABLE 1.10 (continued)

Application/material	Consumption (1000 tons)
Plumbing	
Acrylic	11
Polyacetal	9
Polyester, thermoplastic	3
Polystyrene	2
Reinforced polyester[b]	55
Resin-bonded woods	
Phenolic	195
Urea and melamine	386
Vapor barriers	
LDPE	80
PVC[d]	18
Wall coverings	
PS	3
PVC	57
Total	3159

[a] Exluding bonding or adhesive materials.
[b] Including reinforcements.
[c] Including windows, rainwater systems, etc.
[d] Including swimming pool liners.

Reprinted from Ref. 3, courtesy McGraw-Hill.

the major building material of the twenty-first century. By 1985 almost 4 million tons of plastics will be consumed by this market.

PVC is the major plastic used in construction. If the use of PVC grows as expected, so will the consumption of typical PVC additives, such as plasticizers and stabilizers. The main growth of PVC products is in sidings, roofing, and cellular interior trim.

Because of recent developments in ABS, such as foam-core pipe and the use of ABS in water well casings, the market for ABS is expected to grow until about 1985.

Acrylics are also expanding their markets, as their advantages over glass are becoming better known.

In recent years the rigid plastic foams market has grown 15 to 20% per year. Because of the current need for energy conservation, the consumption of rigid foams is expected to continue to grow.

Epoxies in construction are growing by about 10% per year, mainly in flooring, bridge and road construction, and coatings. A similar growth is predicted for acetal homopolymers and co-polymers, mainly in plumbing [3].

B. Packaging

Since packaging constitutes a considerable proportion of total solid waste (in 1976 about 35% of the municipal and 3.5% of industrial waste [11]), this market for plastics must be discussed in some detail.

The packaging industry as a whole constitutes one of the largest manufacturing groups in the United States. Sales of the packaging industry in 1976 were estimated to be $32.6 billion (about 3% of the value of all finished goods sold in the United States [11]. Table 1.11 shows values of sales of various products by the packaging industry. It can be seen that plastics and paper occupy third place, each accounting for 12% of the market.

Virtually every known plastic material is used in some aspect of packaging: films, coatings, containers and lids, closures, and adhesives. Of the 15.4 million tons of plastics consumed in the United States in 1978, over 3.56 million tons (approximately 23%) found their way into some form of package or packaging material. Table 1.12 shows major applications for plastics in packaging.

As shown by the figures, packaging constitutes the largest market for some polymers while some others are used only marginally. The advantages offered by plastics in terms of energy savings, ease of handling, and appearance assure the future growth of plastics usage at the cost of other more traditional materials.

TABLE 1.11 Sales of the U.S. Packaging Industry

Material	Value ($ billion)		
	1970	1977	1980
Paperboard	6.2	11.7	13.1
Metals	5.1	9.9	12.5
Plastics	1.8	4.3	5.8
Paper	2.5	4.1	4.9
Glass	1.9	3.2	4.1
Miscellaneous	1.0	1.9	2.1

Based on data from Ref. 4.

1. CELLULOSICS IN PACKAGING

The main application for cellulosics is in blister packaging, but they also find applications in transparent rigid containers, windows in cardboard boxes, displays, vials, bottles, and closures.

Cellulose derivatives used in blister packaging are: cellulose acetate, cellulose acetate butyrate, and cellulose acetate propionate. These materials are used to produce films with good transparency for an attractive product display. Approximately 19,000 tons of various cellulosics were used in packaging applications in 1977 [12].

2. ETHYLENE VINYL ACETATE IN PACKAGING

Although ethylene vinyl acetate was one of the earliest packaging plastics, it still finds numerous applications in that area. Ethylene vinyl acetate is used in such diverse applications as pallet stretch wrap film, adhesives, and coextruded film or sheet [12].

TABLE 1.12 Consumption of Plastics in Packaging, 1978

Material	Consumption (1000 tons)	Material	Consumption (1000 tons)	Material	Consumption (1000 tons)
Adhesives		Containers and lids		PS, thermoformed	
Polyvinyl acetate	27	ABS	11	Foam	87
Polyvinyl alcohol	8	Cellulosics	12	Impact	143
Other	7	Thermoformed	8	Oriented	80
Total	42	Other		Other	54
		HDPE		PVC	
Coatings (for paper, film, foil, metal)		Blow-molded, up to 2 gal	494	Blow-molded	46
EVA copolymer	46	Blow-molded, 2 gal or more	60	Thermoformed (including blister packs)	34
HDPE	19	Injection-molded	118	SAN, molded	7
DPE	219	Thermoformed	25	Thermoplastic Polyester (PET)	86
PP	7	LDPE		Urethane foam	31
PVA	21	Blow-molded	25	Other	28
PVC	10	Injection-molded	75	Total	1700
Other	43	PP		Film	
Total	365	Blow-molded	28	HDPE	70
		Extruded (drinking straws)	15	LDPE	973
Closures		Injection-molded	32	PP	109
Phenolic	9	Thermoformed	7	PS	20
HDPE	37	PS, blow-molded	15	PVC	82
LDPE	14	PS, molded		Other	40
PP	55	Crystal	67	Total	1294
PS	30	Foam	56		
PVC (gaskets and liners)	11	Impact	56	Grand total	3564
Urea	7				
Total	163				

Introduction to Plastics

3. EXPANDED POLYSTYRENE IN PACKAGING

Expanded polystyrene (EPS) shows faster growth in the packaging field than any other material. For the past 10 years it has grown at the pace of 8% per year, which is more than twice the growth of corrugated paperboard.

This growth rate does not appear to be decreasing. The major reason for the present and projected growth is a trend toward packaging ever larger parts with EPS. The use of EPS packages for heavy automotive parts, electronic components, appliances, and so on, as well as innovations in EPS package design, are ensuring continuous market growth of this product [12].

4. HIGH-DENSITY POLYETHYLENE IN PACKAGING

HDPE has the properties required for blow molding, and an excellent impact strength. Consequently, it is extensively used to manufacture bottles for milk, bleaches, detergents, antifreeze, and so on, as well as for other food containers, caps, closures, and beverage cases. Because of its properties, over 800,000 tons of HDPE were used in packaging in 1978, which is over 50% of the total HDPE production. In the packaging field, HDPE competes with paper, glass, wood, metal, and other materials. The annual growth of HDPE in packaging is 10 to 12%, which compares favorably with the overall growth of the packaging market (approximately 3% per year [12].

5. LOW-DENSITY POLYETHYLENE IN PACKAGING

Packaging film constitutes the largest portion of the LDPE market in packaging. LDPE film is characterized by a desirable combination of flexibility, toughness, clarity, chemical inertness, and low cost, the properties which caused LDPE to capture about 70% of the packaging film market and grow at an annual rate of 10 to 13% [12].

6. *POLYPROPYLENE IN PACKAGING*

Polypropylene is a relatively young and versatile resin, showing the fastest growth rate of all commodity resins. As new functions and applications for PP are being developed, its growth potential continues to increase. Economics also favor growth of PP in packaging. The price differential between propylene and ethylene is continuing to increase. Because of its low density, PP has the lowest cost per unit volume of all plastic materials.

The largest single market for PP is films, the industry consuming over 100,000 tons in 1978. With the development of stretch blow molding, the process capable of improving contact clarity and mechanical properties, the PP bottle market is growing rapidly.

The high-speed vacuum-forming process developed by European and American manufacturers is increasing the use of PP in such applications as margarine and dairy tubs, drinking cups, food service trays, and other thin-walled containers. It is predicted that this portion of the market will be consuming over 200,000 tons of PP resin per year by 1981 [12].

7. *SOLID POLYSTYRENE IN PACKAGING*

Solid polystyrene is used mainly for closures and for containers and lids. Although solid polystyrene constitutes only about 12% of the plastics used in packaging, it captures over 25% of the markets in which it competes. Approximately two-thirds of polystyrene used in packaging is of the high-impact grade, and the remaining one-third is general-purpose grades [12].

8. *POLYVINYL CHLORIDE IN PACKAGING*

Polyvinyl chloride (PVC) is probably the most versatile plastic used by the packaging industry. Its properties include good gas

Introduction to Plastics

and moisture barrier characteristics, high clarity, easy colorability, ability to accept various surface finishes (for example, gloss or matte), heat sealability, and lack of odor and flavor. PVC can also be formulated to meet specific requirements, such as specified tear-resistant properties, and resistance to UV transmittance and UV degradation.

9. THERMOPLASTIC POLYESTER IN PACKAGING

This resin (polyethylene terephthalate or PET) has been used since the late 1950s when it was introduced for vegetable boiling bags. In spite of a rather long history, its real growth started in 1978 when its use in soft drink bottles was first initiated. Such bottles require good mechanical properties, good CO_2 barrier characteristics, chemical purity, and excellent clarity. This balance of properties and ease of manufacturing are characteristic of the resin. PET soft drink bottles are becoming the fastest selling new item in the history of the packaging industry, and the total packaging market for PET is predicted to reach over 200,000 tons by the early 1980s. The material will be used in such applications as packages for toiletries and cosmetics, health-care products, household chemicals, beverages, and foods [12].

10. POLYBUTYLENE IN PACKAGING

Polybutylene is not yet widely used in packaging, but because of its properties it is expected to capture a considerable portion of the packaging film market in the near future. Polybutylene films have an excellent toughness: about 3 times the impact strength and up to 8 times the tear strength of LDPE films. Polybutylene films retain a greater percentage of their strength at elevated temperatures, which makes them suitable for hot-fill applications [12].

C. Transportation

In 1985 the 27.5-mpg fleet average takes effect as stipulated by the Department of Transportation. It is estimated that this average will be partially achieved by using 350 to 600 lb of plastics per car [13].

Plastics first penetrated the automotive market as materials used in vehicle interiors. Vinyl-coated fabrics were used for automotive seats since the late 1940s. In the early 1970s rigid plastic interior components such as instrument panels, defogger grilles, vents, and so on, replaced similar components made of metal. Average 1978 model car interiors used approximately 98 lb of plastics. The largest portion of that total is polyurethane seating foam ($\sim$24 lb). The use of plastics in the car exterior is relatively new. Because of the advantages of low weight, corrosion resistance, and impact-absorbing characteristics, the 1978 model car used somewhat over 100 lb of plastics in exterior applications. They are used for bumpers, fender liners, front end panels, grilles, window louvers, and so on. Under-the-hood applications of plastics account for only $\sim$32 lb in the 1978 model. Plastics are being used for ducts, battery cases, battery trays, and electronic ignition components [14].

The reduction in gasoline consumption can be achieved by reducing the weight of the car, which in turn can be achieved by further substitution of plastics for more traditional materials. The use of reaction injection molding (RIM) will permit a growth in large polyurethane automotive parts. The evolution of material technology will be toward shorter production cycles of polyurethane formulations and stiffer parts. Reinforced polyester, mainly in the form of sheet molding compounds (SMC), follows polyurethane in the quantity used in 1978 cars. The use of SMC will further increase if the problems of surface pinholes and shrink marks are solved. Polypropylene and PVC are the two thermoplastics most

TABLE 1.13 Plastics in Transportation

	1979 (1000 Tons)	
Material	Passenger cars, model year	Other, calendar year
ABS	70	5
Acrylic	21	14
Nylon	25	4
Phenolic	24	3
PP	160	9
Polyurethane	185	21
PVC	128	17
Reinforced polyester	180	19
Other	41	5
Total	834	97

Note: Data for passenger cars are for 1979 model year, data for other transportation are for 1979 calendar year.
Reprinted from Ref. 3, courtesy McGraw-Hill.

widely used in the automobiles. Table 1.13 shows the main markets for plastics in transportation.

D. Furniture

The important factor in the increasing use of plastics in furniture is the more efficient manufacturing techniques possible. Flexible polyurethane dominates the furniture market, followed by PVC and polystyrene. Table 1.14 shows the consumption pattern for plastics in furniture.

TABLE 1.15 Use of Plastics by the Electrical and Electronics Industries (excluding Automotive), 1978

Material	Consumption (1000 Tons)
ABS	14.0
Cellulosics[a]	0.7
Epoxy (electrical laminates)	10.2
Nylon (including wire and cable)	17.1
Phenolic	69.0
Polyacetal	1.5
PC	15.0
Polyester, reinforced	59.0
Polyester, thermoplastic	7.0
HPPE[b]	49.0
LDPE[b]	195.0
PP[b]	7.0
PS	127.0
PVC[b]	183.0
Urea	19.8
Others	7.5
Total	781.8

[a] Excludes appliance uses.
[b] Wire and cable.

Reprinted from Ref. 3, courtesy McGraw-Hill.

F. Housewares

Traditionally, engineering resins accounted for only a small portion of this market. With the advent of microwave ovens the need for specialized plastics dishes opened up a new market for such resins as polysulfones, polycarbonate, and polyester. Table 1.16 shows the 1978 usage of plastics in housewares.

G. Appliances

The pressure to upgrade the quality of products, conform with government standards, and keep down manufacturing costs results in the continuing growth of plastics in appliances. The volumes of various materials used in this market in 1978 are shown in Table 1.17.

TABLE 1.16 Plastics in Housewares, 1978

Material	Consumption (1000 Tons)
Melamine	20
Phenolic	21
HDPE	144
LDPE	200
PD	42
PS	175
PVC	25
Styrene acrylonitrile	9
Other	19
Total	655

Reprinted from Ref. 3, courtesy McGraw-Hill.

Introduction to Plastics

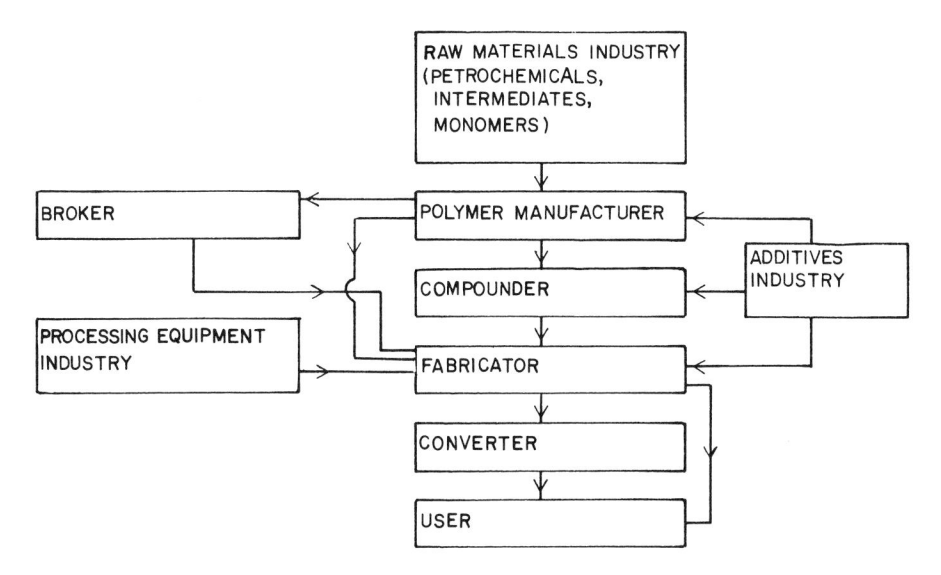

FIG. 1.21 Structure of the plastics industry

A number of companies of various sizes supply processing equipment to the plastics converters. Just as the resin manufacturers play an important role in the development of new resins or the improvement of existing ones, so the equipment manufacturers play an important role in the development of new manufacturing technologies.

VIII. PRICES AND SUPPLIES

Figure 1.23 shows a typical life history of a product. Initial small-scale production (A) shows increasing growth (B) as the new applications are found and the manufacturing technology is being improved. Eventually, the market is becoming saturated (C) and growth of production tapers off to the slope comparable with the overall growth of the economy (D). When better products are developed, the original product looses its markets and finally disappears from the marketplace (E).

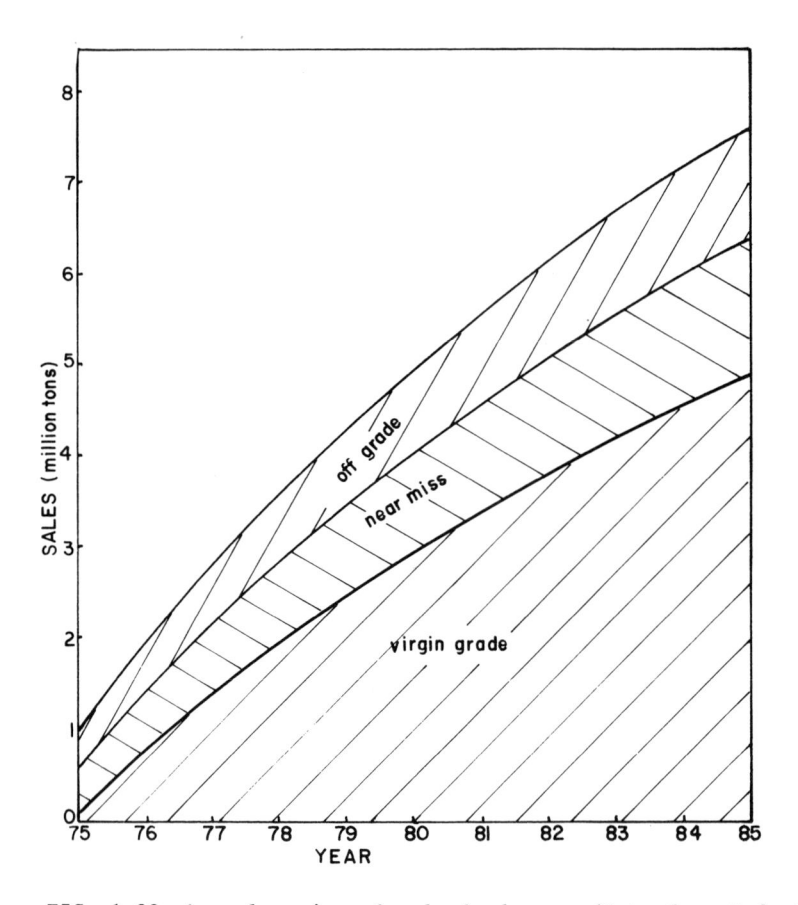

FIG. 1.22 Annual resin sales by brokers. (Data from Ref. 6.)

Out of the major commercial plastics only cellulose nitrate has completed its life history (although it is still being used in rather small quantities). Some plastics appear to have saturated the market, but most of the commodity resins are still in a growth phase and the newer engineering resins are still growing at an accelerating pace (B) [17].

The history of resin prices is probably as interesting as the history of plastics themselves. In the late 1960s the production of plastics was exceeding the market demand. Since the market

Introduction to Plastics   59

was growing rapidly, the resin suppliers felt that inventory
building was necessary for the controlled growth of sales. The
result of this was a decrease in the average price of the resins
at a time when the price of other commodities was growing (Fig.
1.24). In 1972 the demand started catching up with supply. That
year the growth of consumption exceeded the growth of production
by 55%. As early as the first quarter of 1973, there were signs
of resin shortages. At first there was a shortage of PVC followed
by shortages of PS and PE. There are two reasons for the shortages:
(1) the low profit margins in the late sixties and early seventies
did not justify investments in the new production capacities, and
(2) the surprisingly high increase in demand. Suddenly the
prices of plastics started to climb. The problems were compounded
by the oil crisis and the world-wide recession of 1974. The sudden
drop in demand in 1974 did not result in a drop of the prices. On
the contrary, because of the sharp increase in oil prices, the
prices of plastics kept increasing, practically doubling within
2 years. This situation finally focused the attention of the
industry on its dependence on the uncertain supply of raw materials

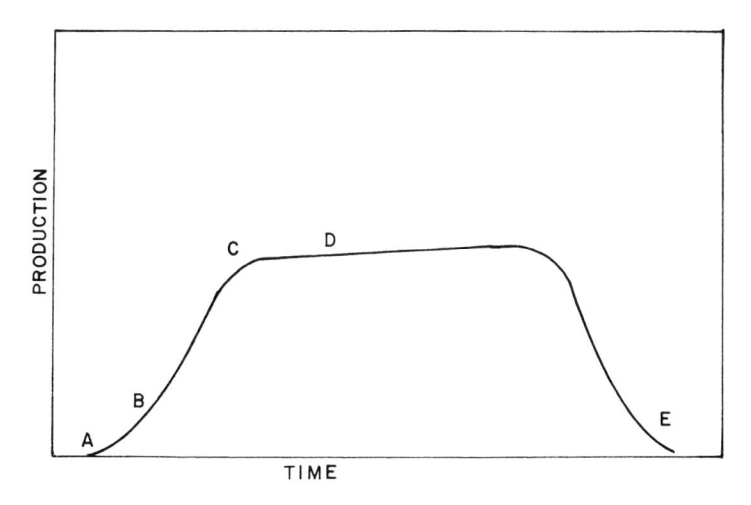

FIG. 1.23 Life cycle of a product. (Reprinted from Ref. 17,
courtesy Society of Plastics Engineers, Inc.)

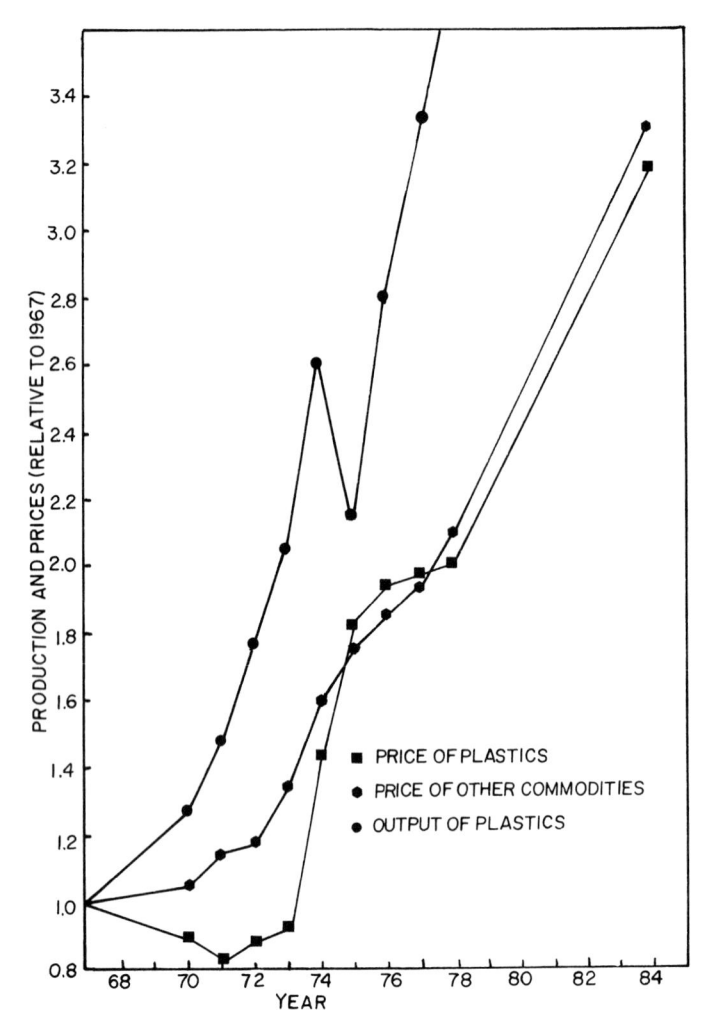

FIG. 1.24 Production and prices of resins. (Based on data from *Mod. Plast.*, Jan. 1975, Jan. 1976, Jan. 1977, Jan. 1978, Jan. 1979, Oct. 1979.)

and the need for proper management of the resources. It is interesting to note that despite the rapid increase in the price of resins (an increase exceeding that of any other commodity), the demand for plastics recovered soon and so did their consumption.

Introduction to Plastics

In 1979 the consumption of plastics showed a healthy growth that is expected to continue. Prices of plastics depend on the price of petroleum feedstock but also on labor costs, costs of construction materials, cost of borrowing money, and so on.

Figure 1.25 shows the effect of the threefold increase of the price of the crude oil on the prices in the downstream operation. The threefold increase caused only 30 to 40% increase in the cost of plastics products.

The short-term prices of plastics will continue to escalate depending on the supply/demand conditions, and the long-term price increase will probably be similar to that of other commodities, that is, 6 to 10% per year. The approximate 1979 prices of some resins are listed in Table 1.18.

It is predicted that the supplies of resins in the 1980s will be adequate but less abundant than current supplies. The following factors will prevent the suppliers from overexpanding resin capacity: (1) cost of new plant construction, (2) difficulty of expanding already large volume production, (3) uncertainty regarding oil supplies, regulations inflation, and the growth of the economy, and (4) market strategy shifting from expanding volumes to maximizing profits. Because of the general tightness of supply volume, plastics will be susceptible to shortages [19].

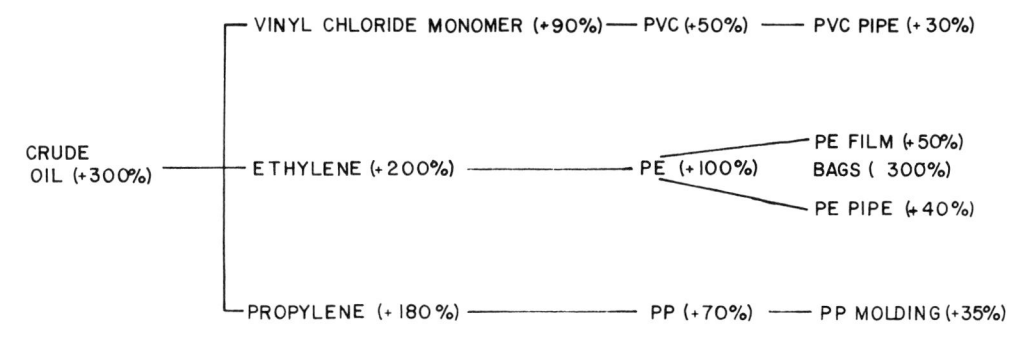

```
                    ┌─ VINYL CHLORIDE MONOMER (+90%) ── PVC (+50%) ──── PVC PIPE (+30%)
                    │
                    │                                              ┌─ PE FILM (+50%)
CRUDE               │                                              │
OIL (+300%) ────────┼─ ETHYLENE (+200%) ──────── PE (+100%)  BAGS ( 300%)
                    │                                              │
                    │                                              └─ PE PIPE (+40%)
                    │
                    └─ PROPYLENE (+180%) ──────────── PP (+70%) ──── PP MOLDING (+35%)
```

FIG. 1.25 The effect of increased oil prices on downstream operations. (Reprinted with permission from Ref. 18, copyright 1976, Pergamon Press.)

TABLE 1.18 Prices of Selected Resins, 1978

Material	Price (cents/lb)	Material	Price (cents/lb)
ABS		HDPE	
Low-impact	47½	Blow molding	30¼-31¼
Medium-impact	57½	Extrusion	31-3/4
High-impact	64	Injection molding	30¼-31¼
Acrylic molding powder	61	Wire and cable	32½
Cellulosics		LDPE	
Acetate	87	Liner-grade	28½-31½
Butyrate	89	Clarity film	30½-33½
Propionate	89	Injection molding	29½-31½
Epoxy	81	PP	
Melamine molding powder	55-60	Homopolymer	30
Nylon-6	116	Medium-impact copolymer	33-35½
		High-impact copolymer	33-36½
Nylon-6/6	116	PS	
Nylon-11	233	Crystal	28-29
Phenolic molding compound	40	High impact	29-30
Polyacetal	100	Polyurethane	
PC	113	Polymeric MDI	61-63
		Polymeric TDI	57-59
Polyesters		Flexible slabstock polyol	38
Unsaturated (thermoset)	37-38	Thermoplastic PUR	150
Thermoplastic PBT	110	PVC	
Thermoplastic PET	47	Homopolymer dispersion	41-43
		Homopolymer suspension	28-29
		Styrene acrylonitrile (SAN)	45
		Urea molding powder	56

REFERENCES

1. J. A. Brydson, *Plastics Materials,* Butterworth, London, 1975.
2. "For an Energy-Efficient America Petrochemical and Plastics Feedstocks Are Vital," *SPI*, May 1979.
3. "Materials 79," *Mod. Plast. 56(1)*, 1979.
4. "Resin Scene: It's Go to 1980," *Plast. Eng. 31(8)*, 1975.
5. "Resin Report 78," *Plast. Eng. 34(9)*, 1978.
6. *Modern Plastics Encyclopedia, 1978-79*, McGraw-Hill, New York, 1978.
7. "Chemicals and Additives, Opening the Door to New Markets," *Mod. Plast. 55(9)*, 1978.
8. R. B. Saymour, "The Role of Fillers and Reinforcements in Plastics Technology," *Polym. Plast. Technol, Eng. 7(1)*, 1976.
9. "Goodby Resin Shortage," *Mod. Plast. 52(1),* 1975.
10. "Plastics in Construction: Coming on Strong in 1978," *Plast. Eng. 34(7)*, 1978.
11. C. H. Kline, Kline Guide to the Packaging Industry 1977, *Industrial Marketing Guide JMG-9-27*
12. "Plastics in Packaging," *Plast. Eng. 34(10)*, 1978.
13. "What Detroit Wants from Plastics and Vice Versa," *Mod. Plast. 55(5),* 1978.
14. "Facts and Figures of the Plastics Industry," *SPI*, 1978.
15. J. Milgrom, *Incentives for Recycling and Reuse of Plastics*, Report for U.S. Environmental Protection Agency, Arthur D. Little, Inc., Cambridge, Mass., 1972.
16. "Now Broker Is a Respectable Word," *Mod. Plast. 55(1)*, 1978.
17. R. D. Deanin and C. C. Kourkoulakas, "Predicting Future of Plastic Production," *33rd ANTEC SPE*, 1975.
18. W. Kaminsky, J. Menzel, and H. Sinn, "Recycling of Plastics," *Cons. Rec. (1)*, 1976.
19. "Blueprint 80's," *Mod. Plast. 56(5)*, 1979.

Chapter 2

SOURCES OF PLASTICS WASTE

I. DEFINITIONS REGARDING PLASTICS WASTE

Before the generation of plastics waste can be discussed, one must define some important terms related to the topic. Milgrom [1] suggests a relatively detailed system for the classification of plastics waste. That system takes into account ease of recycling, product category, physical category, and production segment in which the waste has been generated, and so on. The simple classification system used in this work is contained in the following definitions:

Waste plastics consist of plastics resin or product that must be reprocessed or disposed of.

Industrial plastics waste is a plastics waste generated by various industrial sectors.

Postconsumer plastics waste is a plastics waste generated by a consumer.

Nuisance plastics (NP) are waste plastics that cannot be reprocessed under the existing technoeconomic conditions.

Scrap plastics (SP) are waste plastics that are capable of being reprocessed into commercially acceptable plastic products.

Primary recycling is the processing of scrap plastic into the same or similar types of product from which it has been generated, using standard plastics processing methods.

Secondary recycling is the reprocessing of scrap plastic into plastic products with less demanding properties.
Tertiary recycling is the recovery of chemicals from waste plastics.
Quarternary recycling is the recovery of energy from waste plastics.

The distinction between nuisance plastics and scrap plastics is a rather vague one. It depends on the economic and technological conditions prevailing at a given time. Some plastics classified as nuisance plastics a few years ago (for example, some phenolic resin wastes) are being recycled now and must be reclassified as scrap plastics. The division between primary and secondary recycling processes is often quite arbitrary. Recycling of phenolics has been classified as secondary, although the material is often added to the virgin compound and used to produce the same product with minimal change in properties. The reason for classifying the process as secondary recycling is that recycled phenolic mixed with a virgin compound acts only as nonreactive filler, and by itself does not have the properties of the virgin compound. Similar problems might be encountered in distinguishing between tertiary and quarternary recycling. Some incineration plants utilize a two-step process in which the waste is first pyrolyzed and the products of the pyrolysis incinerated. Although the process is used to recover energy, the fact that the waste is first converted into a different type of fuel classifies it as tertiary recycling.

II. PLASTICS CYCLE

The flow of plastics products as well as plastics waste is shown schematically in Fig. 2.1. The resin manufacturer supplies the fabricator and compounder with raw materials. Scrap plastics from the resin manufacturer are sold to the reprocessor for reprocessing, or to the fabricator as second-grade resins.

The fabricator sells his product to the converter, packager, assembler, or consumer. Scrap plastics are recycled within the

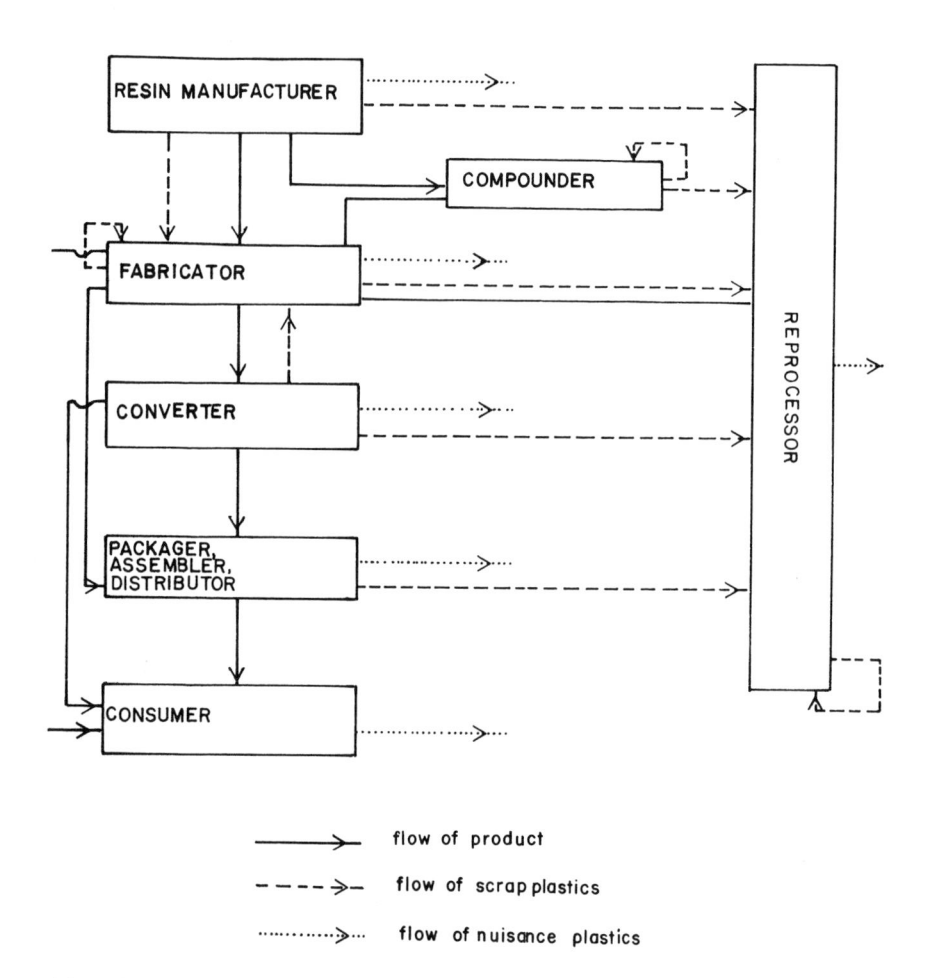

FIG. 2.1 Flow of plastics products and plastics waste.

fabricator's plant or sold to a reprocessor. Raw materials can come from the resin supplier or compounder (virgin material), reprocessor (reprocessed scrap), or converter (scrap plastics).

The compounder buys resins from the resin manufacturer, compounds them with additives, and resells them to the fabricator. The scrap plastics can be reprocessed in the compounder's own plant or sold to the reprocessor.

Sources of Plastics Waste

The converter purchases a semifinished plastic product (e.g., plastic film) and converts it into the finished product (e.g., a plastic package). The plastics scrap generated by the converter goes to the reprocessor or fabricator. The packager, assembler, and distributor purchase finished products from the fabricator or converter, assemble and package them, and sell them to the consumer, though not necessarily directly. This segment generates nuisance plastics, but virtually no scrap plastics.

The reprocessor purchases scrap plastic from various industrial sectors, reprocesses them, and sells them to the fabricator. Small amounts of scrap plastics generated by the reprocessor are recycled in plant.

The consumer, at the end of the plastics cycle, generates only nuisance plastics, which eventually end up together with nuisance plastics generated by the industrial sector in landfill or an incinerator.

III. GENERATION OF INDUSTRIAL PLASTICS WASTE

Although a considerable amount of data is available regarding postconsumer plastics waste, very little is known about industrial plastics waste. Industrial plastics waste generated in the United States will reach a volume of approximately 3 million tons by 1980 and this amount will more than triple by the year 2000. Presently, approximately 4.7% of total plastics production is recycled by reprocessors and over 8% is recycled in plant [1]. Industrial plastics scrap, being reprocessed by the industry and thus often considered an "avoidance of waste," is usually omitted by those researching the subject.

The data presented here are derived from Refs. 1-5, with future predictions based on an assumption of proportionality between the amount of industrial plastics waste and total plastics production. This assumption can be used to obtain only very rough estimates, since there is currently a trend toward the development of scrapless plastics processing equipment resulting in the decreasing generation of industrial plastics waste by some manufacturing sectors.

A. Plastics Waste Generated by Resin Manufacturer

Table 2.1 gives approximate amounts of plastics waste generated by the manufacturers of commodity resins. The amount of waste depends on such factors as the complexity of the polymerization process, the number of grades produced in a single plant, and the number of manufacturing steps. The least waste is generated in the manufacturing of polyethylene, the most in the manufacturing of PVC.

Plastics waste is generated during the polymerization process (reactor scrappings, unsuccessful runs, etc.), compounding (extruder purgings, unsuccessful runs), and shipment and storage (Table 2.2).

B. Plastics Waste Generated by Fabricator

Table 2.3 shows the approximate amounts of plastics waste generated by fabricators from commodity resins.

Most of the waste plastics generated by blow molding come from the pinchoff. Some products with a large percentage of pinchoff (e.g., bottles with handles) can generate up to 40% waste. Waste plastics are also generated during equipment setup and purging.

Waste plastics generated by injection molding come from sprues and runners (routinely reground and added to the virgin resin in amounts of 5 to 20%), machine setup, and machine purging.

Extrusion processes generate waste during equipment start-up, from purging and trimming, and cutting of the final product.

Waste plastics are generated in the calendering process in the form of drippings from the mixers and the calender, and from the trimmings and product rejects.

Rotomolding generates plastics waste in the form of trim flash at mold parting lines, removed openings in the product, and product rejects.

Sources of Plastics Waste

TABLE 2.1 Plastics Waste Generated by Resin Producer

Resin	% Waste plastics[a]		Estimated amount of waste plastics (U.S.) (million tons)[b]					
			NP			SP		
	NP	SP	1978	1980	2000	1978	1980	2000
PE	0.9	2.5	0.054	0.078	0.240	0.150	0.220	0.680
PP	2.2	4.0	0.035	0.059	0.275	0.064	0.110	0.500
PS	0.9	5.0	0.026	0.035	0.111	0.146	0.195	0.620
PVC	3.0	7.0	0.093	0.120	0.889	0.217	0.280	0.508
Total	1.5	4.0	0.208	0.292	1.515	0.577	0.805	2.308

[a]From Ref. 1.
[b]Author's estimate.

TABLE 2.2 Plastics Waste Generated in Various Manufacturing Steps by Resin Producers

	% Waste plastics	
Process	NP[a]	SP[b]
Polymerization	1.1	1
Adding additives and colorants	0.2	2
Preparation for shipment	0.15	0
Storage	0.05	0

[a] From Ref. 1.
[b] Author's estimate.

A large portion of industrial plastics waste generated by the fabricator is in a relatively clean, chemically uniform state, and can be easily reprocessed. Only about one-third of the plastics waste generated by that segment is classified as nuisance plastic.

C. Plastics Waste Generated by Compounder and Reprocessor

The compounder and reprocessor account for only small percentages of overall plastics waste. Both generate waste by purging of the compounding equipment and unsuccessful runs (off specific products). About 10% of the total throughput of these two segments ends up as plastic waste, a large portion of it classified as scrap plastic.

D. Plastics Waste Generated by Converter

Table 2.4 shows the approximate breakdown of plastics waste generated by a converter using commodity resins. The converter purchases semifinished plastic products from the fabricator and converts them into finished products. Since the converter usually deals with a number of semifinished products (often already decorated), the waste (in the form of trim, or rejected products)

Sources of Plastics Waste

TABLE 2.3 Plastics Waste Generated by Fabricator in Selected Processes (Commodity Resins Only)

Process	% Commodity[b] resins affected	% Waste plastics[a]		Estimated amount of waste plastics[b] (U.S.) (million tons)					
				NP			SP		
		NP	SP	1978	1980	2000	1978	1980	2000
Blow molding	8.9	1.4	3.0	0.020	0.024	0.081	0.042	0.052	0.172
Injection molding	25.7	2.0	10.0	0.080	0.100	0.332	0.410	0.496	1.660
Extrusion (film and sheet)	17.0	1.1	2.0	0.030	0.036	0.121	0.054	0.066	0.022
Extrusion coating	3.3	6.0	0.5	0.020	0.038	0.128	0.003	0.003	0.011
Extrusion (wire and cable)	6.5	5.2	3.0	0.054	0.065	0.218	0.031	0.038	0.126
Extrusion (profile and pipe)	10.3	2.2	3.0	0.036	0.044	0.146	0.049	0.060	0.200
Coextrusion	0.8	9.9	0.0	0.012	0.014	0.047	0.000	0.000	0.000
Calendering	5.0	1.0	2.0	0.008	0.010	0.032	0.016	0.019	0.065
Rotomolding	6.7	3.3	1.0	0.055	0.043	0.143	0.011	0.013	0.043
Total	84.2	2.4	4.7	0.277	0.331	1.248	0.605	0.747	2.299

[a] From Ref. 1.
[b] Author's estimate.

TABLE 2.4 Plastics Waste Generated by Converter in Selected Processes (Commodity Resins Only)

Operation	% Commodity[b] resins affected	% Waste plastics		Estimated amount of waste plastics (U.S.) (million tons)[b]					
		NP[a]	SP[b]	NP			SP		
				1978	1980	2000	1978	1980	2000
Decorating 3-D products	5.6	0.3	0.3	0.003	0.004	0.122	0.003	0.004	0.012
Decorating film	6.7	4.0	2.0	0.043	0.052	0.175	0.022	0.026	0.089
Sealing bags	12.3	3.9	2.0	0.076	0.091	0.309	0.039	0.047	0.158
Cutting	8.5	10.0	10.0	0.135	0.162	0.548	0.135	0.162	0.548
Thermoforming	7.0	3.0	6.0	0.033	0.040	0.134	0.017	0.020	0.069
Total	40.1	4.8	5.0	0.290	0.348	1.288	0.216	0.259	0.876

[a]From Ref. 1.
[b]Author's estimate.

is more difficult to reprocess than that generated by the fabricator. Relatively clean and uniform scrap plastic (for example, trim from thermoformed parts) can be returned to the fabricator for regrinding and addition to virgin resin. Some other scrap plastics (for example, decorated reject cups) can be sold to the reprocessor to be pelletized and sold as a second-grade resin. Less than 50% of waste plastics generated by this segment are suitable for reprocessing.

E. Plastics Waste Generated by Packager, Assembler, and Distributor

Table 2.5 shows the amounts of plastic waste generated by the packager, assembler, and distributor. The waste generated by this sector is characterized by a considerable contamination with non-plastic matter and thus is unsuitable for reprocessing.

IV. PLASTICS IN SOLID WASTE

A. Composition of Solid Refuse

The composition of refuse can be estimated using input or output analysis. Input analysis utilizes production statistics of various products to calculate their amounts in the refuse. An example of such a calculation is given by Nagda and Babcock [6]. Let us assume that Fig. 2.2a shows the sale volume of certain products (S_N) in the year N. The distribution of the lifetimes of these products is given in Fig. 2.2b. The first peak represents the fraction of the products with the 1-year average life span and the second peak corresponds to the remaining products with the longer life span. The distribution of the life spans can be simplified by reduction of the continuous distribution into two discrete outputs (Fig. 2.2c). T (weighted average life span) may be calculated from the distribution of the life spans of various product categories excluding short-lived discardable period D_N.

TABLE 2.5 Plastics Waste Generated by Packager, Assembler, and Distributor (Commodity Resin Only)

Operation	% Commodity[a] resins affected	% Waste plastics[a] NP	Estimated amount of waste plastics (U.S.) (million tons)[b]		
			NP		
			2978	1980	2000
Assembly; packaging	68	0.7	0.076	0.092	0.307
Storage, distribution, displaying	68	0.5	0.108	0.066	0.220
Total	68	1.2	0.184	0.158	0.527

[a]Based on Ref. 1
[b]Author's estimate.

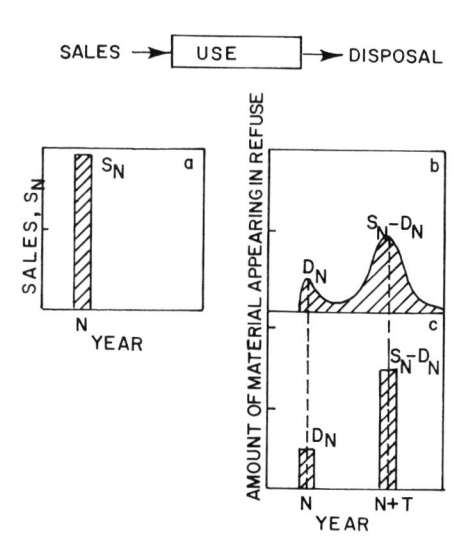

FIG. 2.2 Example of input analysis: (a) Production volume in year N. (b) Life-span distribution. (c) Simplified life-span distribution. (Reprinted from Ref. 6, courtesy Rodale Press, Inc.)

$$T = \frac{\sum_{i=1}^{n} (t_i M_i)}{\sum_{i=1}^{n} M_i} \quad (1)$$

where t_i is the average lifetime of category i, M_i is the quantity of category i, and n is the number of categories (excluding D_N). The total amount of material (Q) appearing in the refuse in the year N + T is

$$Q_{(N+T)} = D_{(N+T)} + S_{(N+T)} \quad (2)$$

The example of the input analysis developed in [6] and shown in Eqs. (1) and (2) has been designed specifically to predict the amount of plastics in refuse. Similar analysis may be carried out for other materials. More complex life-span distributions may also be used. The analysis shown here assumes that at the end of its lifetime the product is disposed and not recycled.

TABLE 2.6 Input Analysis of U.S. Solid Waste, 1971

Municipal solid waste stream (×1000 tons)	Quantity per material balance analyses of domestic and commercial sources of wastes (1000 tons)		
	Inputs	Recycled	Refuse
Paper			
Newsprint	9,800	2,200	7,600
Corrugated	14,100	4,100	10,000
Other	26,900	6,300	20,000
Total	50,800	12,600	38,200
Glass			
Containers	10,000	300	9,700
Other	1,600	500	1,100
Total	11,600	800	10,800
Ferrous metals			
Containers and closures	5,800	400	5,400
Domestic and commercial durables	3,800	300	3,500
Total	9,600	700	8,900
Aluminum			
Cans and closures	310	60	250
Foil	200	--	200
Consumer durables	200	20	180
Total	710	80	630
Tin			
Tinplate	31	3	28
Foil	1	--	1
Other	58	21	37
Total	90	24	66
Copper			
Total production	1,772	1,417	355
Lead	30	--	30
Textiles			
Total production	4,000	1,100	2,900
Rubber			
Total production	2,600	1,000	1,600

TABLE 2.6 (continued)

Municipal solid waste stream (×1000 tons)	Quantity per material balance analyses of domestic and commercial sources of wastes (1000 tons)		
	Inputs	Recycled	Refuse
Plastics			
Containers and film	1,900	--	1,900
Toys	200	--	200
Consumer appliances and durables	100	--	100
Total	2,200	--	2,200
Food, animal, plant, and other wastes			49,319
Total			115,000

Note: For comments and notes see Ref. 7.
Reprinted from Ref. 7, courtesy National Center for Resource Recovery.

The amount of material in the solid waste may also be calculated if the consumption, the amount recycled, and the amount in the liquid waste are known. For example, in 1971 59 million tons of paper were consumed in the United States. Of this amount, 13 million tons were recycled and 8 million tons absorbed by the liquid waste stream or diverted to permanent uses; 38 million tons of paper ended up in the solid waste stream. Table 2.6 contains an example of the results of such calculations.

Output analysis is a common method for calculating the composition of the refuse. It involves hand sorting the garbage and weighing the components. Some problems might be encountered in this method because of the moisture transfer when certain hydrophilic materials (for example, paper) come into contact with moisture-containing products (for example, foodstuffs). Table 2.7 compares the results of input and output analysis of U.S. solid waste (approximately, in 1971). The results are usually fairly

TABLE 2.7 Comparison of Input and Output Analysis

Material	Quantity per input analysis Refuse (1000 tons)	% Total	Quantity per input analysis Refuse (1000 tons)	% Total
Paper	38,200	33.2	42,435	36.9
Glass	10,800	9.4	9,775	8.5
Ferrous	8,900	7.7	7,475	6.5
Aluminum	630	0.55	920	0.8
Tin	66	0.06	58	0.05
Copper	355	0.3	184	0.16
Lead	30	0.03	20	0.0017
Textiles	2,900	2.5	2,185	1.9
Rubber	1,600	1.4	805	0.7
Plastics	2,200	1.9	1,265	1.1
Food, animal, plant, and other wastes	49,319	42.9	49,878	43.4
Total	115,000	99.9	115,000	100.0

Reprinted from Ref. 7, courtesy National Center for Resource Recovery.

close, although some discrepancies may be caused by inaccuracies in statistical data used in the input analysis or by too small a sample size in the output analysis.

Vaughan et al. [8] estimated the total amount of solid waste generated in the United States to the year 2000 (Table 2.8). The analysis assumes a constant waste generation rate per capita for 1970 and 1975 followed by a slight decrease due to conservation, recycling, and government regulations. The total annual amount of solid waste is expected to increase from 187 million tons in 1970 to 244 million tons by the year 2000. The amount of collected solid waste is expected to increase during the same time period from 125 million tons to 222 million tons.

TABLE 2.8 Forecast of Population and Amount of Solid Waste to the Year 2000

	1970	1975	1980	1985	1990	1995	2000
Population							
Total ($\times 10^6$)	205	217	232	249	265	280	297
Farm ($\times 10^6$)[a]	9.8	10.0	10.3	10.7	11.0	11.5	11.9
Urban ($\times 10^6$)	143.5	161.6	179.7	197.7	215.8	233.9	252.0
Rural ($\times 10^6$)	51.7	45.4	42.0	40.6	38.2	34.6	33.1
Solid waste generated							
Pounds per capita per day (a)	5	5	4.9	4.8	4.7	4.6	4.5
Total, $\times 10^6$ tons/year	187	198	207	218	227	235	244
Municipal + rural, 10^6 tons/year	178	185	198	209	218	225	234
Solid waste collected							
Percent of municipal and rural	70	75	80	85	90	95	95
Amount, $\times 10^6$ tons/year	125	139	159	177	196	214	222

Reprinted from Ref. 8.

B. Plastics in the Disposal Area

Plastics constitute presently (1980) only 2 to 3% of municipal refuse. This amount is expected to increase gradually, reaching perhaps 10 to 15% by the year 2000. Milgrom [9] estimated the amounts of plastics in the disposal area (Table 2.9). Polyolefins constitute the majority (70%) of plastics in the disposal area followed by styrene polymers and PVC. Table 2.10 gives typical life spans of various categories of plastics products. Those with a life span of 0 to 1 year are estimated to constitute approximately

TABLE 2.9 Plastics in the Disposal Area, 1970

Type of plastic	From packaging weight		From all sources weight	
	(Million lb)	(%)	(Million lb)	(%)
Polyolefins	3240	82.6	4231	70.6
Styrene polymers	445	11.3	1006	16.8
PVC	240	6.1	755	12.6
Total	3925	100	5992	100

Reprinted from Ref. 9, courtesy New Science Publications.
This table first appeared in *New Scientist*, London, the weekly review of Science and Technology.)

67% of the solid refuse, those with the life span 1 to 5 years approximately 20% [1].

Table 2.11 gives an estimate of the composition and the amounts of plastic products in the solid refuse. As expected, packaging waste constitutes the majority (60%) of the plastic components of the solid waste.

Although plastics constitute only a small portion of the solid waste, on the national scale the total amount of plastic in the refuse is staggering. If one assumes that the plastic waste has the value of only 10 cents per pound (as opposed to approximately 60 cents per pound for similar virgin resins), the total value of plastic waste present in the disposal area in the United States in 1980 is approximately $1.9 billion.

V. FUTURE OF WASTE DISPOSAL

For many years incineration and landfill were the main, if not the only, methods of solid waste disposal. Because of the drive toward energy and materials conservation, as well as the concern about the environments, new waste disposal options are being developed and implemented. Vaughan et al. [8] projected the amounts of wastes disposed by various methods up to the year 2000 (Fig. 2.3). The improved collection procedures and more stringent government

TABLE 2.10 Typical Life Spans of Plastics Products

Product	Estimated life (years)
Production loss	0
Packaging	<1
Novelties	<1
Photographic film	<1
Disposables (dinnerware, hospital goods)	<1
Construction sheeting	2
Footwear	2
Apparel	4
Household goods	5
Toys	5
Jewelry	5
Sporting goods	7
Automotive	10
Gramophone records	10
Luggage	10
Appliances	10
Furniture	10
Cameras	10
Wire and cable	15
Business machines	15
Miscellaneous electrical equipment	15
Hardware	15
Instruments	15
Magnetic tape	15
Construction	25

Reprinted from Ref. 9, courtesy New Science Publications.)
(This table first appeared in *New Scientist*, London, the weekly review of Science and Technology.)

legislation will result in an increase in the percentage of collected waste (close to 90% expected by the year 2000). Open dump and open burning will be drastically reduced, becoming practically insignificant by the year 2000. Sanitary landfill

TABLE 2.11 Plastics Products in the Disposal Area

Type of product	1970 weight (Million lb)	(%)	1975 weight (Million lb)	(%)	1980 weight (Million lb)	(%)
Packaging	3,925	60.1	6,445	56.2	10,170	54.0
Industrial wastes	1,000	15.3	1,830	15.9	3,050	16.2
Housewares	425	6.5	885	7.7	1,270	6.7
Toys	310	4.7	555	4.8	945	5.0
Construction and agricultural sheeting	130	2.0	195	1.7	285	1.5
Appliances	100	1.5	230	2.0	440	2.3
Novelties, disposables	100	1.5	200	1.7	400	2.1
Footwear	90	1.4	140	1.2	190	1.0
Records	95	1.4	140	1.2	205	1.1
Transport	90	1.4	250	2.2	470	2.5
Furniture	60	0.9	170	1.5	355	1.9
Construction	50	0.8	100	0.9	150	0.8
Wire and cable	40	0.6	95	0.8	480	2.5
Others	120	1.8	230	2.0	430	2.3
Total	6,535		11,465		18,840	

will be the main method of waste disposal (40%), but resource recovery and thermal treatment (incineration and pyrolysis) will become economically significant disposal methods. Since by the year 2000 approximately 30% of solid refuse will be recycled, large quantities of waste plastics (possibly up to 1 million tons a year) will become available for processing into secondary products.

Sources of Plastics Waste

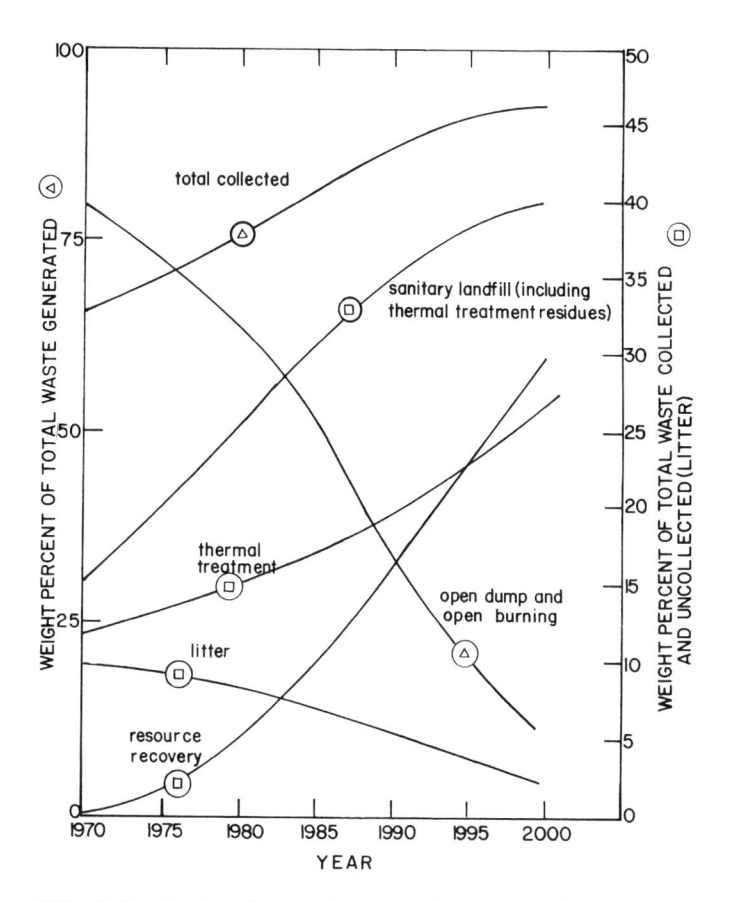

FIG. 2.3 Projections of waste disposed of by various disposal systems. (Reprinted from Ref. 8.)

REFERENCES

1. J. Milgrom, *Incentives for Recycling and Reuse of Plastics*, Report No. EPA-SW-41C-72, Arthur D. Little, Inc., Cambridge, Mass., 1972.

2. J. Milgrom, "Recycling of Plastics," 11th National State-of-the Art Symposium, Division of Industrial and Engineering Chemistry, American Chemical Society, Washington, D.C., June 1975.

3. A. J. Warner, C. H. Parker, and B. Baum, "Solid Waste Management of Plastics," Report for the Manufacturing Chemists Association, De Bell & Richardson, Inc. Enfield, Conn., 1970.

4. C. W. Marynowski, *Disposal of Polymer Solid Wastes by Primary Polymer Producers and Plastics Fabricators*, Report for the U.S. Environmental Protection Agency, Stanford Research Institute, Stanford, Calif., 1972.

5. M. Sittig, *Pollution Control in the Plastics and Rubber Industry*, Noyes Data Corp., Park Ridge, N.J., 1975.

6. N. L. Nagda and L. R. Babcock, "Use of Plastics: A Growing Problem in Solid Waste Disposal??" *Compost Science*, March-April 1973.

7. "Municipal Solid Waste . . . Its Volume, Composition and Value," *NCRR Bulletin 3(2)*, 1973.

8. D. A. Vaughan, C. Ifeadi, R. A. Markle, and H. H. Krause, *Environmental Assessment of Future Disposal Methods for Plastics in Municipal Solid Waste*, Report No. PB-243 366 for the U.S. Environmental Protection Agency, Battelle Columbus Laboratories, Columbus, Ohio, 1971.

9. J. Milgrom, "Identifying the Nuisance Plastics," *New Scientist, 57(830)*, 1973.

Chapter 3

SEPARATION

I. SEPARATION OF COMPONENTS OF MUNICIPAL REFUSE

Municipal refuse can be treated as a potential source of raw materials. Not only do the individual constituents have some economic value, but the municipal refuse, if it must be disposed of without the recovery of value, actually has negative value. The scarcity of raw materials and the concern about environmentally safe disposal of refuse have added a new dimension to the question of recycling solid wastes. Although plastics constitute only a small portion of municipal refuse, the actual quantity is enormous. In order to recycle the components of municipal refuse, they must first be separated; numerous commercial and experimental processes are available for this purpose. Most of the existing separation plants consist of two main stages: preparation of the feed (size reduction), and separation.

A. Size Reduction

Size reduction of municipal solid refuse is the mechanical separation of the material into smaller pieces. It is accomplished mainly by the mechanical forces of tension, compression, and shear applied by crushers, shears, shredders, chippers, rasp mills, drum pulverizers, disk mills, pulpers, and hammermills. Figure 3.1 schematically shows the four main types of crushers. All of these

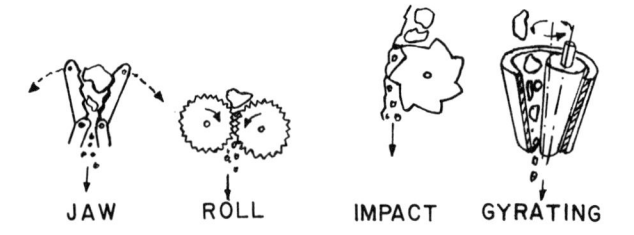

FIG. 3.1 Main types of crushers. (Reprinted from Ref. 1.)

may be used in solid waste separation plants, although impact crushers have the widest application. Jaw, roll, and gyrating crushers are used mainly for brittle materials.

The single-bladed shear is shown in Fig. 3.2; a slow application of shear forces is used to cut bulky materials. Shredders and chippers (Fig. 3.3) utilize both tensile and shearing force. The cutting type is not widely used in the size reduction of municipal refuse because of the vulnerability of the blades to damage. The "pierce and tear" types use toothed wheels rotating at different speeds, which penetrate, shear, and shred the material. This type of shredder is useful in the reduction of ductile or fibrous materials, and is often used on the paper and fiber portion of municipal solid refuse. Rasp mills (Fig. 3.4) use tension, compression, and shear forces. The rotor, fitted on a vertical shaft, carries heavy rasping arms (up to 25 ft in diameter) which force the

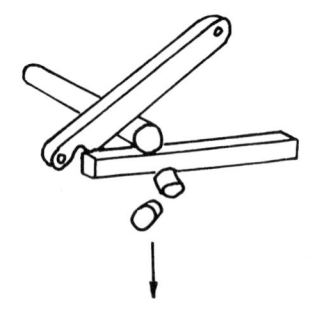

FIG. 3.2 Single alligator-type shears. (Reprinted from Ref. 1.)

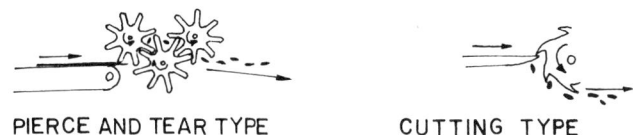

Separation

PIERCE AND TEAR TYPE   **CUTTING TYPE**

FIG. 3.3 Shredders, cutters, and chippers. (Reprinted from Ref. 1.)

waste over rasping pins and through holes in the bottom of the chamber. A reject chute is periodically opened to allow the exit of items too bulky and durable to be properly reduced. Drum pulverizers (Fig. 3.5) have an action similar to the rasp mills. The waste passes through a rotating drum of circular, octagonal, or hexagonal design containing stationary or counterrotating beaters or baffles. The reducible material is torn and pushed through holes in the drum while the nonreducible portion leaves at the lower end.

Figure 3.6 shows a schematic drawing of a disk mill. Such mills consist of a single rotating disk and a fixed contact surface, or two counterrotating high-speed disks. Refuse introduced between the disks is subjected to repeated impact from rotating disk or segmented wheel. Material is reduced in size until it passes through openings in the fixed contact surface. Disk mills are limited to small particle size feed.

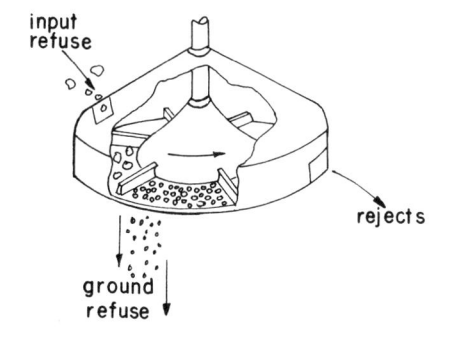

FIG. 3.4 Rasp mills. (Reprinted from Ref. 1.)

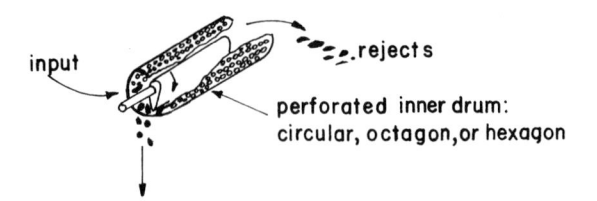

FIG. 3.5 Drum pulverizer. (Reprinted from Ref. 1.)

The wet pulper (Fig. 3.7) is similar to the disk mill. The waste is mixed with water to produce a slurry, which is then introduced into a pulper consisting of a segmented blade rotating at high speed inside a cylindrical housing. The material, reduced to the desired size, exits through openings in the bottom of the pulper. Materials which do not undergo size reduction are rejected ballistically by the rotating blades to the outer portions of the pulper drum, and collected separately.

Hammermills (Fig. 3.8) comprise the largest proportion of municipal solid waste size reduction equipment. They consist of a single or multiple rotor axle with attached hammers, and work by rapidly applying the forces of tension, compression, and shear. There are two general types of hammermills: the swing hammer type (hammers mounted flexibly to the rotor shaft) and the rigid hammer type (hammers mounted rigidly to the rotor shaft). The action necessary to produce size reduction is obtained by positioning fixed blocks on the inside of the hammermill wall. Hammermills may be used for a broad range of feed materials [1,2].

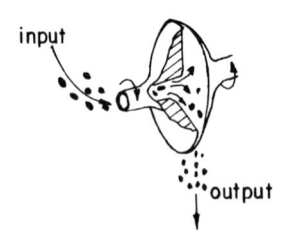

FIG. 3.6 Disk mill. (Reprinted from Ref. 1.)

Separation

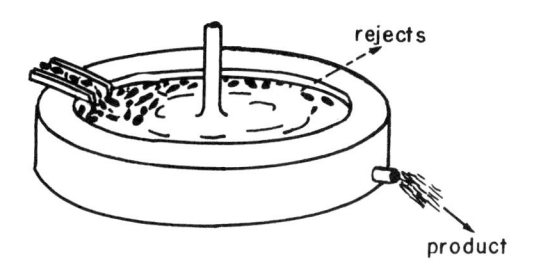

FIG. 3.7 Wet pulper. (Reprinted from Ref. 1.)

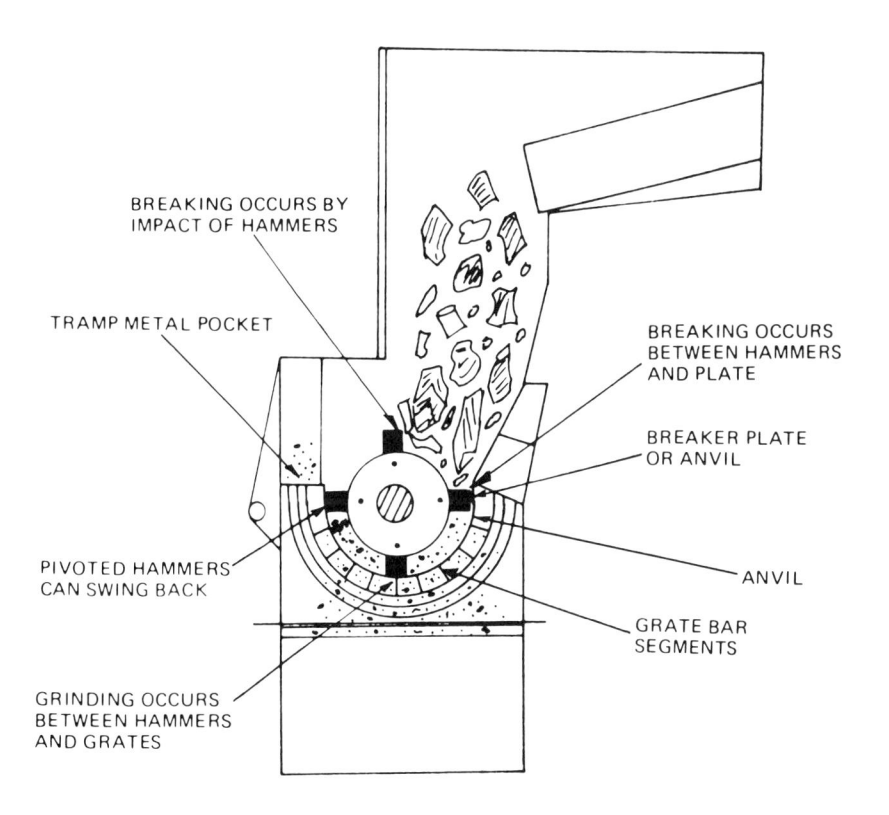

FIG. 3.8 Schematic of hammermill. (From J. L. Pavoni, J. E. Heer, and D. J. Hagerty, *Handbook of Solid Waste Disposal.* Copyright 1975 by Litton Educational Publishing, Inc. Reprinted with permission of Van Nostrand Reinhold Company.)

B. Separation Methods

The methods used to separate the individual components of solid waste are based on differences in their physical properties. The following properties are widely used as a basis for separation processes:

Particle Size. Since such properties as ductility, strength, and impact resistance vary from one material to another, the components of municipal refuse will vary greatly in particle size following the reduction stage. If a given component has a particle size considerably larger or smaller than the others, it can be separated from the rest of the refuse on that basis.

Density. There are a variety of separation methods utilizing density differences.

Electromagnetism. Ferrous metals are recovered from refuse using magnetic separators. The principle is widely used because of its simplicity.

Color. The difference in appearance of various components of solid waste has been a criterion for many years in hand sorting. Automated separation methods based on color differences are not available. They are ideally suited for the segregation of materials such as glass, and for the further separation mixed waste glass into color categories [2].

The preceding fundamental material properties are the basis of the following separation methods:

Manual Separation. Manual separation is the oldest technique still used in less affluent societies. It is of little importance in modern separation plants. The method is sometimes used at the incoming feed belt to scavenge easily separable items. Hand separation of paper, cardboard, glass containers, and so on, is best practiced at the source.

Gravity Separation. Widely used types of equipment for gravity separation are vibrating tables, ballistic separators, inclined conveyors for removal of stones and other heavy particles, and fluidized bed separators.

Air Classification. Three parameters are utilized in air classification: size, specific gravity, and shape. Generally zigzag or similar-shaped columns are used. Air enters at the bottom of the column and the material to be separated in the middle. A series of columns with different geometries or flow rates will produce several grades of materials. Usually the light fractions at the top are collected in cyclone separators while the air and dust are returned to the column. Instead of a vertical column separator, a vortex classifier may be used in which a radially inward flowing air vortex replaces the upward flow. Other types of air classifiers utilize horizontal air flow, rectangular chamber with zigzag baffles, a rotary cylinder and a current of air, and vertical vanes. The choice of a specific system depends on the type of feed and the desired degree of separation. Air classification is the most widely used technique for the separation of solid wastes.

Magnetic Separation. This method is used in virtually all systems for the removal of ferrous metals. Separation is achieved by mechanisms such as magnetic pulleys, dry and wet drum separators, and cross-belt separators.

Electrostatic Separation. Electrostatic separation is somewhat similar to magnetic separation. It relies on the ability of some materials, such as plastics or paper, to acquire and hold an electrostatic charge. The particles are attracted to a charged roll or belt, or are deflected in an electrostatic field.

Color Separation. Optical separation has found an application in the sorting of glass according to color. It may find a similar application in the separation of different colored plastics [3].

C. Materials Recovery Processes

Processes for separating components of solid wastes may be classified as wet or dry. Dry processes are generally simpler, require lower energy input, and have fewer potential environmental problems. Wet processes are usually capable of producing much cleaner and more uniform products. Dry and wet separation techniques may be used as parts of the same process. The Black-Clawson and Fläkt systems, descriptions of which follow, illustrate the principles of the wet and dry processes, respectively.

1. *Black-Clawson Hydrasposal/Fibreclaim Process*

Figure 3.9 shows a flowsheet of the Black-Clawson materials recovery system. Raw waste from trucks is conveyed to a Hydropulper, where it is converted into a 3 to 4% slurry in water. The slurry, containing paper, food waste, plastics, glass, and small metal pieces, is extracted from the bottom of the pulper through small openings. Larger objects are ejected through an opening near the bottom and removed from the system by a continuous bucket elevator. The discharge from the bucket elevator is washed in a rotary drum washer, and the ferrous metals are removed by magnetic separation. The slurry from the pulper is pumped through a liquid cyclone to remove inorganic materials. The reject residue contains about 80% glass and some aluminum. The next step is to reduce to discrete fibers any paper that has not disintegrated thus far, and to screen out such materials as plastics, wood, and leather. This is accomplished on a so-called VR Classifier, a heavy screen plate with 1/8-in. diameter perforations, and a high-speed rotor operating against it. The pulsating action of the rotor prevents the screen from plugging. Objects larger than 1/16-in. are removed in a second stage of screening, while small slivers and fine dirt are removed in a second stage of centrifugal cleaning.

After leaving the centrifugal cleaner the slurry is pumped over an inclined screen with horizontal slots. About 85% of the water goes through the slots, carrying with it very short fibers, debris, clay, and food particles. The recovered fibers are dewatered in two stages. The equipment for the first stage is an inclined thickener consisting of a perforated cylinder set at a 60° angle within which rotates a screw. The screw conveys the pulp through the perforated cylinder and discharges 10% pulp into a cone press where it is de-

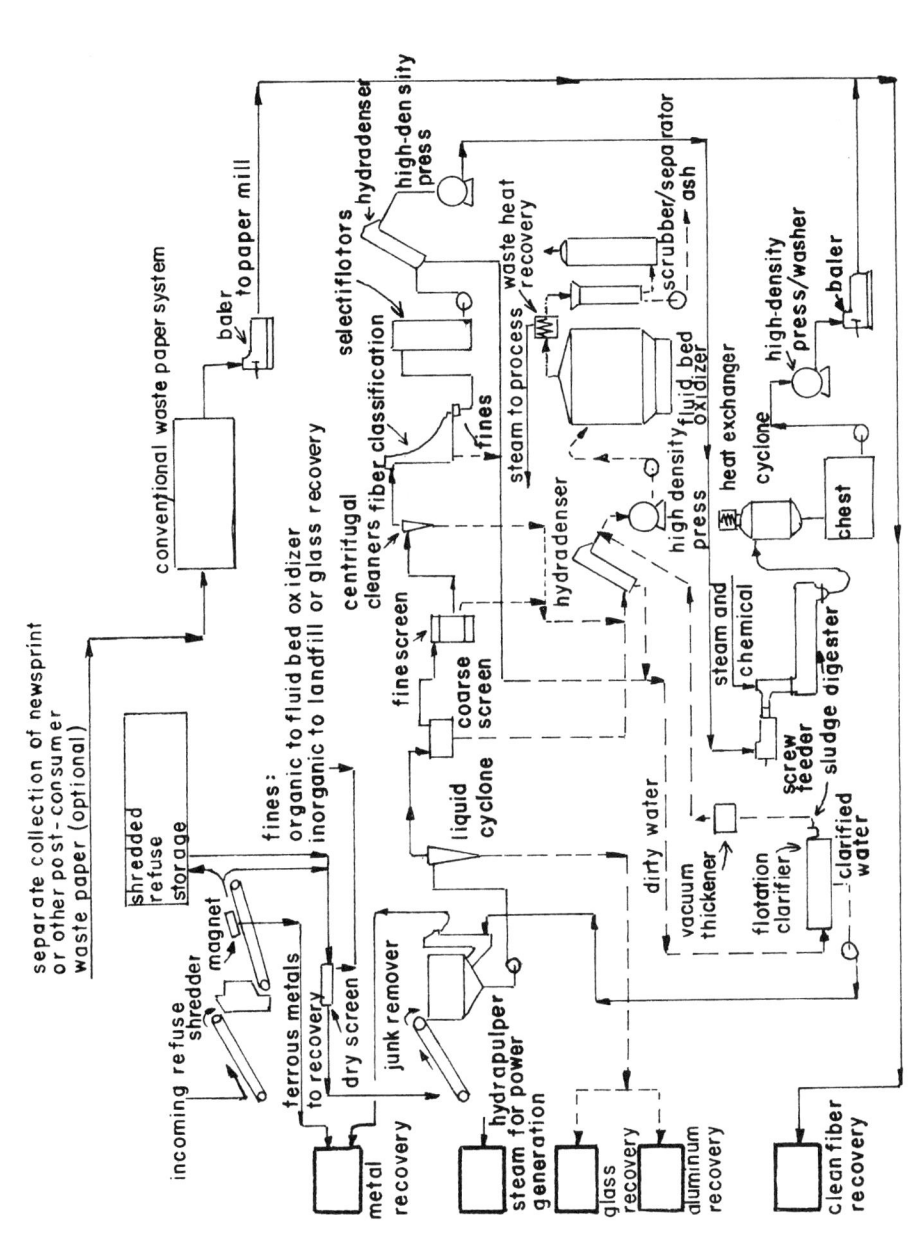

FIG. 3.9 Flowsheet for Black-Clawson Hydrasposal/Fibreclaim system. (Reprinted from Ref. 4, courtesy Noyes Data Corp.)

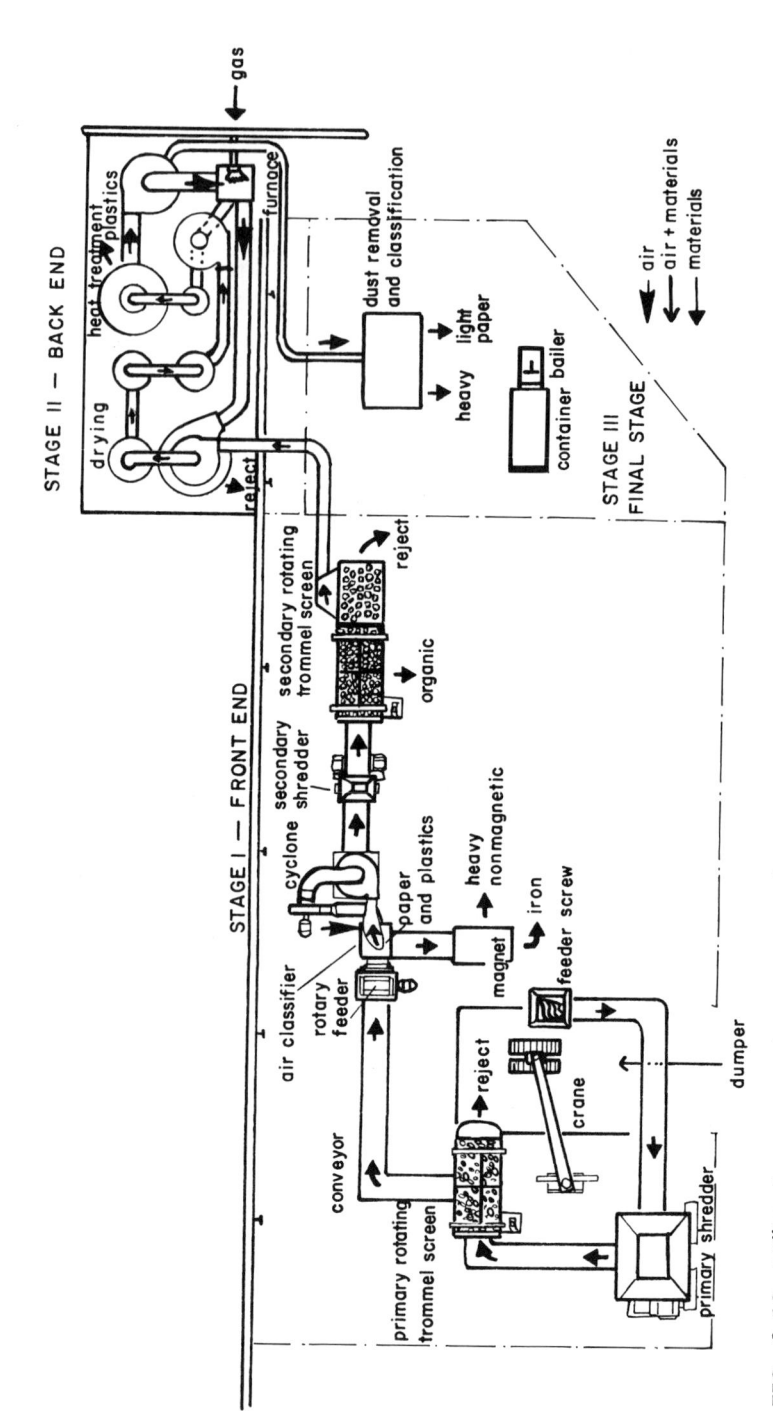

FIG. 3.10 Fläkt dry separation system. (Reprinted from Ref. 5, courtesy AB Svenska Fläktfabriken.)

watered to approximately 40% solids. The main product from the Black-Clawson process is pulp of papermaking quality. The drawbacks of the system are high power demand and the necessity for extensive wastewater treatment [2,4].

2. Fläkt Waste Recovery System

The experimental plant designed and built by AM Svenska Fläktfabriken is a good example of a dry separation system. Figure 3.10 is a schematic of the process. The feedstock is prepared by a combination of shredder (which breaks the plastic bags in which the refuse is packed, thus exposing the material) and a rotating trommel screen (which separates out excessively large items). From the trommel the material is moved by rotary feeder into a zigzag type air classifier. Closed-circuit air circulation is employed and only a small portion of the circulating air is filtered and discharged. The fan motor provides heat, which helps to dry the material being separated.

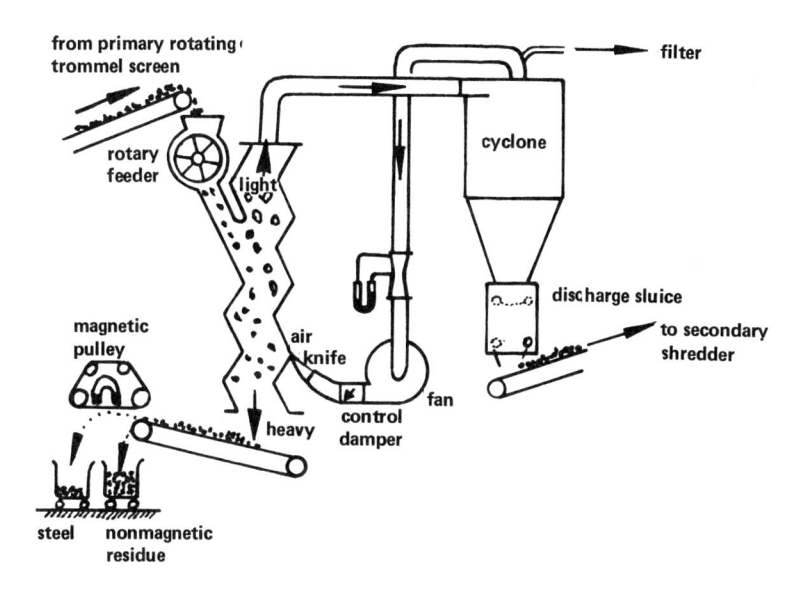

FIG. 3.11 Arrangement of vertical air classifier. (Reprinted from Ref. 5, courtesy AB Svenska Fläktfabriken.)

The light fraction is discharged through the cyclone (Fig. 3.11) onto the rotary feeder and into the secondary shredder. The purpose of the secondary shredder is to liberate the entrained impurities and to achieve a more uniform particle size. The light fraction consists of moist paper, some plastic film, and textiles. The fine impurities, consisting of wood waste, sand, dust, and so on, are separated by the second rotating trommel screen. The pneumatically conveyed material from stage one is a mixture of moist paper partially contaminated by organic matter. Some plastic film and textiles remain.

The purpose of stage two is to carry out further purifications of the paper product and render it biologically stable. This purification is achieved by drying with hot air (Fig. 3.12). The

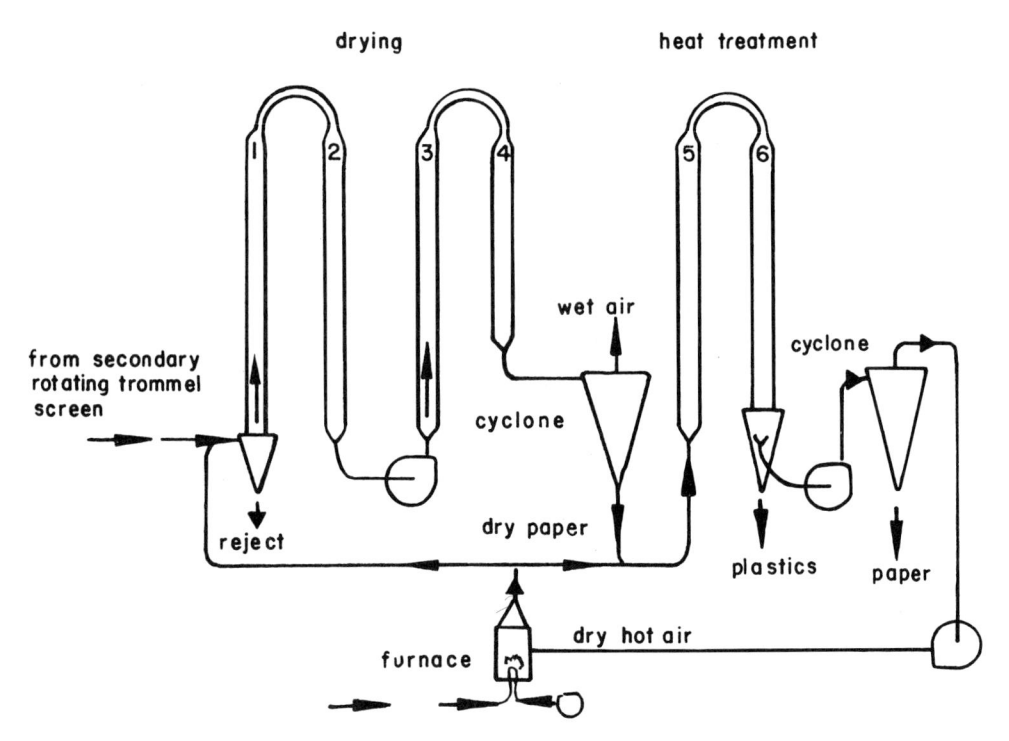

FIG. 3.12 Paper dryer. (Reprinted from Ref. 5, courtesy AB Svenska Fläktfabriken.)

dryer is divided into two stages: in the first the material is dried and the humid air discharged; in the second the material is heat-treated with recycled dry air. The dryer incorporates two separators. In the first heavy material with entrained paper is separated out: the reject consists mainly of textiles. The heat treatment of the second stage is used to destroy bacteria in the paper; plastic is considered an impurity. Although most of the plastic material is separated out by pretreatment in stage one, a certain amount is entrained with the paper. Under the action of heat, the flakes of plastic contract to small balls which can be separated out aerodynamically. This separation is achieved in the last heating tower, which is equipped with a special separator [5].

II. SEPARATION PROCESSES SPECIFIC TO PLASTICS

A. Separation of Paper/Plastics Mixtures

Mixtures of plastics and paper are a common product of municipal waste dry separation plants. To increase the value of the product, the mixture has to be separated into its components. The main problem in separation derives from the aerodynamic similarity of plastic film and paper. Three main principles have been suggested as the basis of separation methods: (1) the application of heat, (2) wet pulping, and (3) electrodynamic separation.

1. *PROCESSES INVOLVING THE APPLICATION OF HEAT*

Two separation processes involving application of heat as a means of changing the properties of the plastic are described by Laundrie and Klungress [6, 7, 8]. Figure 3.13 illustrates the "hot cylinder" method based on a patent by Fyfe [9]. The separation device consists of an electrically heated, chrome-plated cylinder enclosed within a hollow rotating tube fitted with vanes to ensure a tumbling action about the heated cylinder. The drum and the heated cylinder rotate in opposite directions. A doctor blade is in contact with the heated cylinder at the bottom. There is

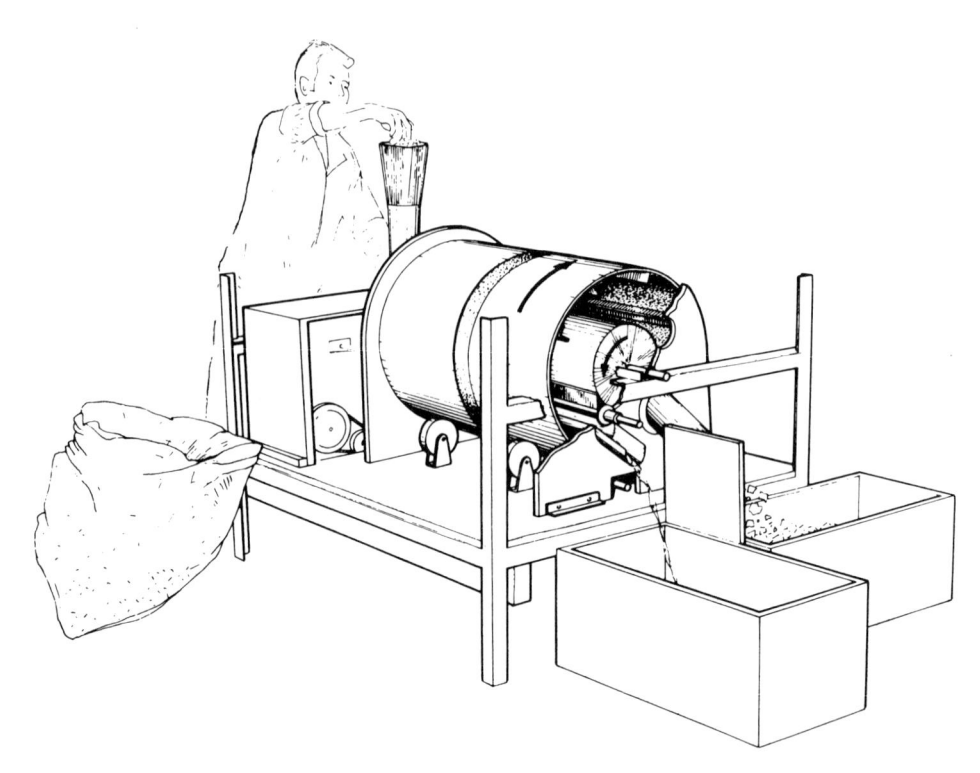

FIG. 3.13 Hot "cylinder" separation method. (Reprinted from Ref. 8, courtesy U.S. Department of Agriculture.)

a trough connected to the bottom of the blade. Material is fed into the drum through a sheet metal tube inserted at the entrance. Plastic materials coming into contact with the hot cylinder melt and are removed by the doctor blade. Table 3.1 shows the efficiency of the process. Removal of over 90% of plastics from the paper can be achieved. The plastic stream is relatively paperfree: 1% or less of paper contaminants can be achieved [6, 8].

Figure 3.14 is a schematic of a process based on the reduction in specific surface area of plastic films due to the action of heat. A mixture of plastics and paper is deposited on the conveyor which transports it into the heating zone. In the system described in Refs. 6-8 an agricultural crop dehydrator is used as the heating

TABLE 3.1 Effectiveness of the "Hot Cylinder" Separation Process

Trial no.	Cylinder speed (rpm)	Agitator speed (rpm)	Cylinder temperature (°F)	Plastic amount (%)	Feed size	Feed moisture (%)	Feed rate (lb/in.)	Efficiency (%)	Paper separated with plastic (%)
1	70	20	365	15	4	10	19	95.6	4.1
2	120	20	365	5	5	10	10	81.6	0.8
3	70	30	365	5	4	Dried	10	57.4, 76.2, 73.1	2.1
4	120	30	365	15	5	Dried	19	38.4	0.7
5	70	20	320	15	5	Dried	10	86.3	1.2
6	120	20	320	5	4	Dried	19	84.2	0
7	70	30	320	5	5	10	19	84.2	1.9
8	120	30	320	15	4	10	10	79.6, 77.2	2.3
9	120	30	320	5	5	Dried	10	54.3	2.0
10	70	30	320	15	4	Dried	19	87.3	1.5
11	120	20	320	15	5	10	19	73.0	2.1
12	70	20	320	5	4	10	10	83.6	0.7
13	120	30	365	5	4	10	19	50.0	1.1
14	70	30	365	15	5	10	10	87.3	2.6
15	120	20	365	15	4	Dried	10	87.3	1.3
16	70	20	365	5	5	Dried	19	89.1	2.1

Reprinted from Ref. 8, courtesy U.S. Department of Agriculture

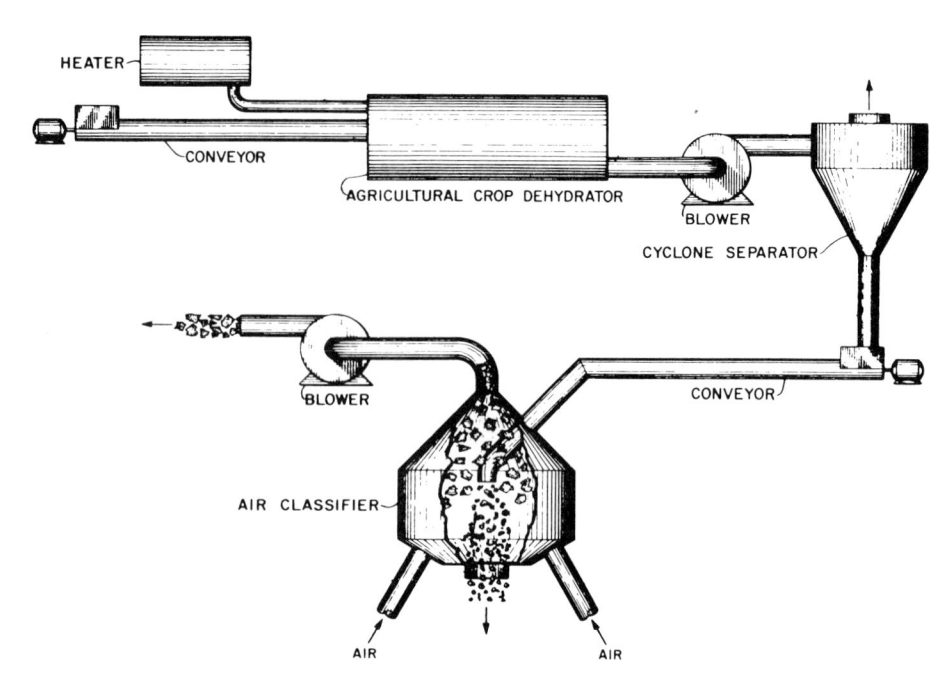

FIG. 3.14   Separation process utilizing specific surface
reduction.   (Reprinted from Ref. 7.)

enclosure. The mixture is transported in a hot gas stream generated by the heater, which causes the thermoplastic film to contract, thus reducing its surface area. The mixture is discharged through the cyclone onto the conveyor, which feeds it into the air separator. Air drawn through the separator carries wastepaper out through the exit duct while thermoplastic particles fall to the bottom of the classifier and are discharged. An almost complete separation of plastics and paper can be achieved using this method [6,7]. A similar process is used in the Fläkt system.

2. WET SEPARATION PROCESS

A wet separation process patented by Black-Clawson Fibreclaim, Inc. [9], is applicable to the recovery of plastics from the light

Separation 101

fraction obtained from the dry separation plant. It can also form
a basis for a complete wet separation process. Figure 3.15 illustrates the application to a dry separation. A conveyor (1)
delivers waste material to a shredder (2). The output from the
shredder is transferred to an air classifier. The light fraction
from the air classifier comprises approximately 60% paper, 20%
plastics, and the balance rags and vegetation residue. The light
fraction is transported to a pulper (3) equipped with rotor (4)
and extraction plate (5) having relatively small perforations.
Pulped paper is capable of passing through the openings. Plastic
particles retained in the tub are discharged from time to time
through a separate outlet (6) provided with a shutoff valve (7).
The plastic-rich fraction is transferred to a dewatering device (8),
and then to an air classifier (9) [10].

3. ELECTRODYNAMIC SEPARATION

Figure 3.16 is the schematic for the electrodynamic separator
illustrated in Fig. 3.17. The mixture of plastics and paper is
fed into the separator by a vibratory feeder (2). The material
falls onto the rotating ground drum (3) and is carried into the
corona formed between the wire beamed electrode (1) and the drum.
The paper is drawn toward the electrode while the plastics adhere
to the drum. As the drum rotates, the plastics are brushed free at
the bottom. Good separation of the 1- to 3-in. shredded material

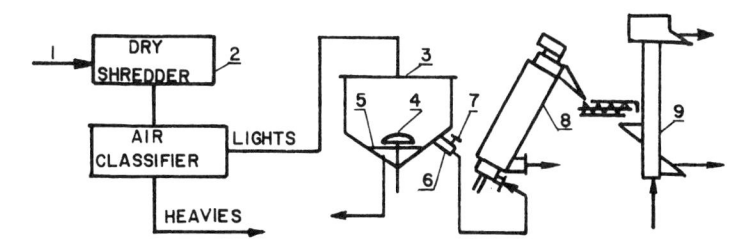

FIG. 3.15 Black-Clawson wet pulping process. (U.S. Pat. 1,512,
257. Reprinted from Ref. 10.)

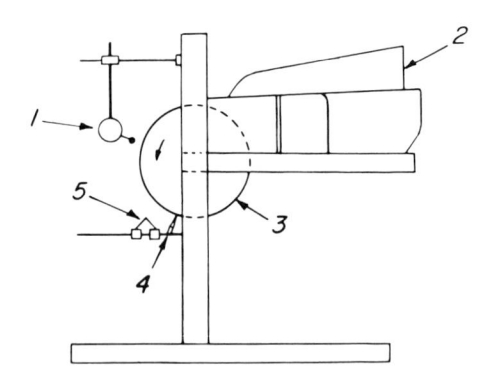

FIG. 3.16 Schematic diagram of electrodynamic separator:
(1) combination 4-in. aluminum electrode with wire electrode;
(2) vibratory feeder; (3) grounded rotating drum; (4) brush;
(5) adjustable stream splitter. (Reprinted from Ref. 11, courtesy
Bureau of Mines, U.S. Department of the Interior.)

FIG. 3.17 Electrodynamic separator used in Bureau of Mines
Experiments. (Reprinted from Ref. 11, courtesy Bureau of Mines,
U.S. Department of the Interior.)

TABLE 3.2 Effect of Moisture Content on the Separation of Plastics from Paper

Moisture (%)	Plastic concentrate (%)		Paper concentrate (%)	
	Plastic	Paper	Paper	Plastic
0	12.0	88.0	0.0	0
10	68.3	31.7	100.0	0
15	72.4	27.6	100.0	0
20	88.1	11.9	100.0	0
25	90.8	9.2	100.0	0
30	94.1	5.9	100.0	0
35	96.5	3.5	100.0	0
40	98.7	1.3	100.0	0
45	99.4	0.6	100.0	0
50	99.8	0.2	100.0	0
55	100.0	0.0	100.0	0

Reprinted from Ref 11, courtesy Bureau of Mines, U.S. Department of Interior.

was obtained at voltages ranging from 35 to 5 kV at a spacing of about 6 in. Table 3.2 shows the effect of moisture on the efficiency of the process. Using feed containing 15% moisture results in a paper fraction free of plastic; however, the plastic material is contaminated with a substantial amount of paper. Increasing the moisture content above 50% results in the complete separation of plastics and paper in one pass [11].

B. Separation of Plastics from Plastic-coated Fabric

A large amount of waste PVC-coated fabric is available, mainly from companies involved in its manufacture or conversion. All

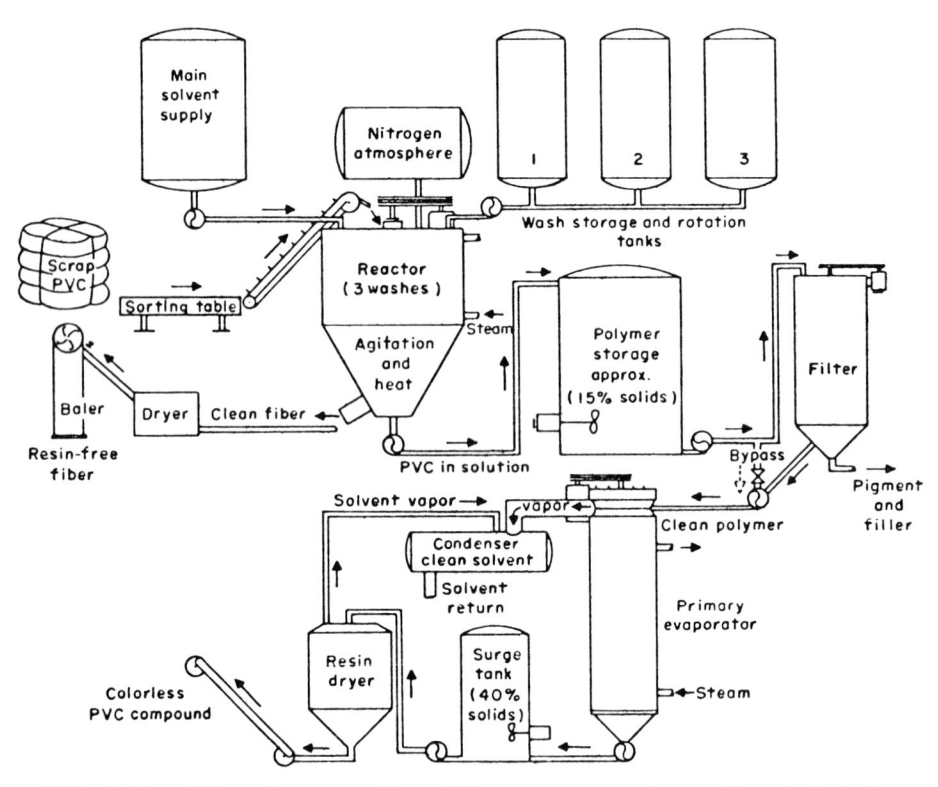

FIG. 3.18 Fiber Process, Inc. PVC recovery system. (Reprinted from Ref. 12, courtesy Society of Plastics Engineers, Inc.)

processes developed for the separation of PVC from fabrics involve solvent extraction of the polymer [12,13].

Figure 3.18 is a flow diagram of a process developed by Fiber Process, Inc. Materials arriving at the plant are cut into sizes suitable for hand sorting of the various colors. The sorted material is dried and loaded into a jacketed vessel, which is then sealed and an inert gas introduced. The vessel is filled with a solvent such as tetrahydrofuran (THF), and the agitated mixture is heated to a temperature slightly below the boiling point of the solvent. The resin dissolves in the solvent, and the solution is transferred to the storage tank. A total of three washes is needed for complete extraction. Solvent trapped in the fibers is driven off by heated nitrogen. The dry fibers are then baled and wrapped for shipment. The polymer solution may be filtered to remove pigments, fillers, or other contaminants, and then fed into a preconcentrator, usually a vertical-film evaporator. The output contains 30 to 40% solids. The final drying takes place under vacuum in a thin-film evaporator or a spray dryer. The dryer yields a colorless, granular PVC resin, or compound of the original formulation. Solvent is condensed and returned to the process. The properties of the recovered resin are the same or better than those of the virgin resin. Since extreme thermal conditions are never introduced, degradation of the resin during separation is avoided. The Fiber Process system recovers the total PVC compound.

The process described by hafner [14] allows recovery of pure polymer and additives; the flowsheet is shown in Fig. 3.19. The solvent is preheated and added to the feed. The mixture is agitated and maintained at an elevated temperature until all PVC has dissolved. The fabric and/or fillers are separated out by filtration and the solution is mixed with a nonsolvent. The nonsolvent does not dissolve PVC, but is completely or almost completely miscible with the solvent; thus the PVC is precipitated. The plasticizers remain in solution; the precipitated polymer is removed by filtration, washed with nonsolvent, and dried. The filtrate contains solvent, nonsolvent, plasticizers, and other additives. The solution is fractionally distilled; the solvent and nonsolvent are separated and returned to the process. The residue containing plasticizers may be directly reused, reused after washing and drying, or first separated into individual components. The process can be used for separating the polymer from PVC-coated fabric or for separating polymer from a compounded PVC [14]. A similar process has been described by Wainer [15], in which water is used as the nonsolvent. In this process the water and solvent must be completely or almost completely miscible in all portions without the formation of azeotropes.

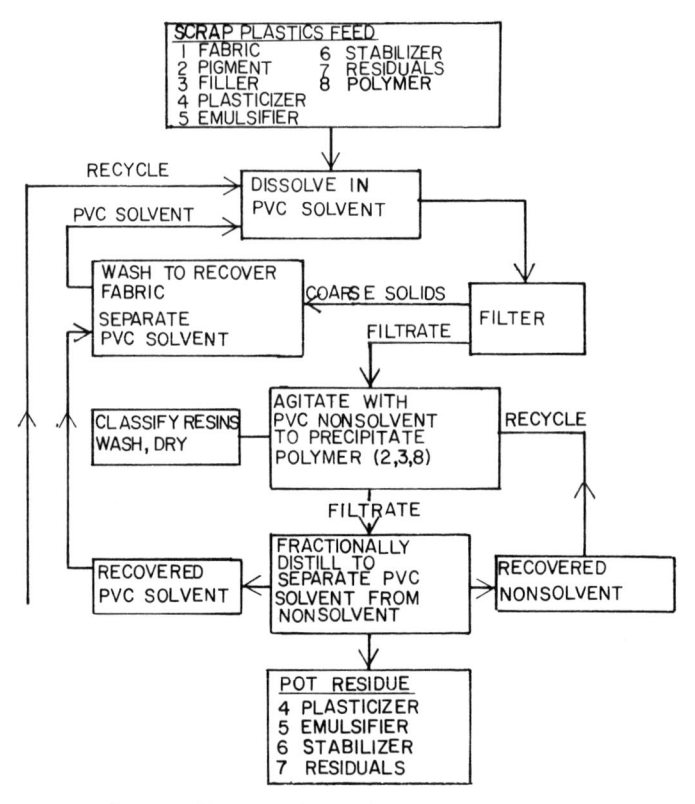

FIG. 3.19 Hafner's vinyl chloride polymer recovery process.
U.S. Pat. 3,836,486. (Reprinted from Ref. 14.)

C. Separation of Polymer from Polymer-coated Wood Fiber

In the 1940s the paper industry began changing over from wax to polymer-coated paper for water-barrier and heat-sealing properties. These widely used polymer coatings now constitute an obstacle to paper recycling efforts. They also constitute a potential source of recycled plastic which could be reused in the same, similar, or less demanding applications. Two separating techniques are described: the wet pulping and solvent extraction processes.

1. WET PULPING PROCESS

Felton [16] describes various wet pulping paper recovery processes. In all these processes the paper is pulped forming an aqueous suspension, while the coating remains in the form of large sheets or particles. Screening can be used to separate pulp from plastic. The pulp is usually relatively clean, but the plastic portion tends to contain some paper fibers.

2. SOLVENT EXTRACTION

The schematic of a solvent recovery system (the Riverside Recovery Process) is given in Fig. 3.20. The process is designed to treat materials such as coated kraft wrapping and packaging papers, plastic-coated food board, oil-saturated kraft, waxed cup stocks, and asphalt-laminated corrugated containers. The separation is achieved by multiple washing of the wastepaper stock with hot solvent (perchloroethylene) and steam stripping the residual solvent from the fiber after the final wash. The process is comprised of the following steps [17]:

Air-dry waste material is charged into the dissolver-dryer.
Solvent extraction is carried out.
Residual solvent is removed from the clean material by steam
    evaporation.
Steam-dry material is ready for pulping.
Steam is condensed and the solvent decanted.
Solvent is distilled, recovered and reused; the coating, adhesive,
    or impregnating polymer is recovered for possible reuse.

D. Separation of Mixtures of Plastics

Both industrial and postconsumer plastics wastes often occur as mixtures of generic groups of plastics. The following separation techniques have been investigated: (1) float/sink methods, (2) processes utilizing differences in surface tension, and (3) solvent extraction. These processes have the potential for separating simple two- or three-component mixtures of industrial plastic wastes, but their usefulness in the separation of a complex postconsumer plastic waste mixture is questionable. Thus far, none of these approaches has been implemented on an industrial scale.

DISSOLVING-DRYING SECTION

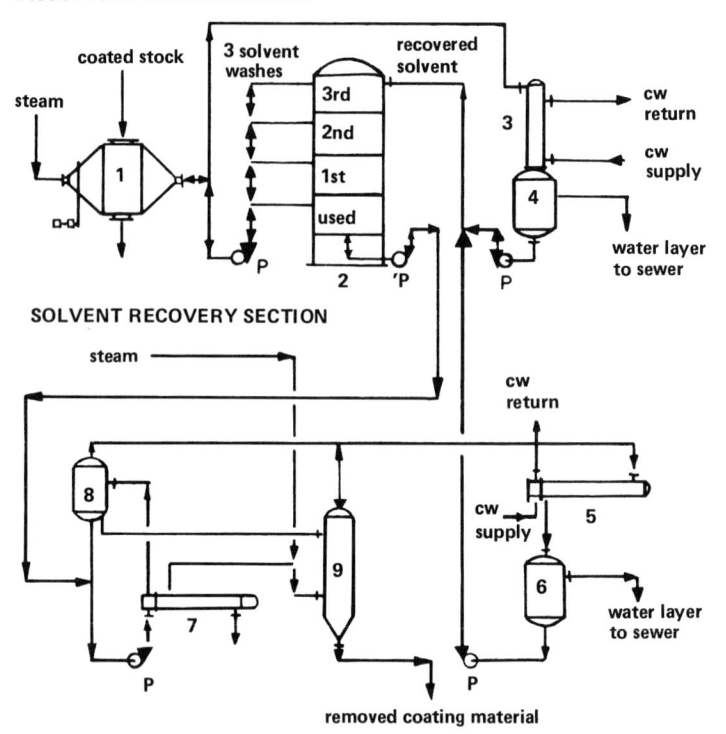

FIG. 3.20 Riverside recovery process: (1) dissolver-dryer; (2) solvent storage tank; (3,5,7) condenser; (4,6) decanter; (8) flash evaporator; (9) finishing still; (9) pump. (From Ref. 17. Reprinted by permission of the publisher.)

1. FLOAT/SINK SEPARATION

The three main plastic components of municipal solid waste-polyolefins, PVC, and PS- have slightly different densities: polyolefins are 0.90 to 0.96, PVC 1.22 to 1.38, and PS 1.05 to 1.06 g/m^3. These density differences can be used to separate a mixture of plastics into generic groups using a sink/float separator. The flowsheet of such a process, proposed by the U.S. Bureau of Mines, is given in Fig. 3.21. The separation is achieved using four

Separation

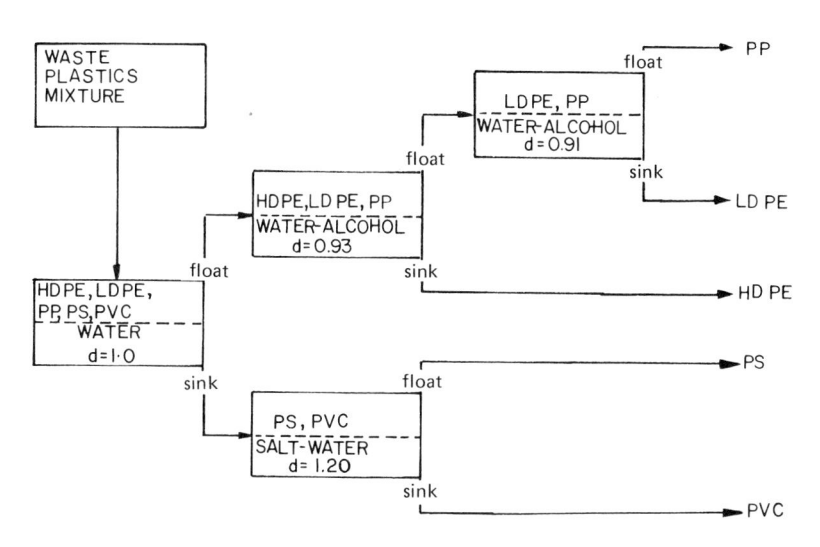

FIG. 3.21 Flowsheet of the sink-float separator. (Reprinted from Ref. 18, courtesy Bureau of Mines, U.S. Department of the Interior.)

liquid media: water (ρ = 1 g/cm^3), two water alcohol mixtures (ρ = 0.93 and 0.91 g/cm^3), and an aqueous salt solution (ρ = 1.20 g/cm^3).

Figure 3.22 shows the schematic of an experimental hydraulic separator using only water as a separating medium: it is a combination of float/sink and elutriation separation. A mixture of chopped plastics is fed into the sink/float separator where the polyolefins are floated off and the other components sink. The heavy fraction is transported to the elutriation column where the PS is carried with the water current and the PVC sinks. The segregated fractions are caught on screens while the water is recycled to the main reservoir. To test the separator, a plastics-rich fraction from Black-Clawson process was upgraded from 14% plastic to an almost all-plastics mixture. Table 3.3 gives the sink/float analysis of this mixture. It can be seen from the table that the polyolefin (PO) and PS fractions are relatively uncontaminated

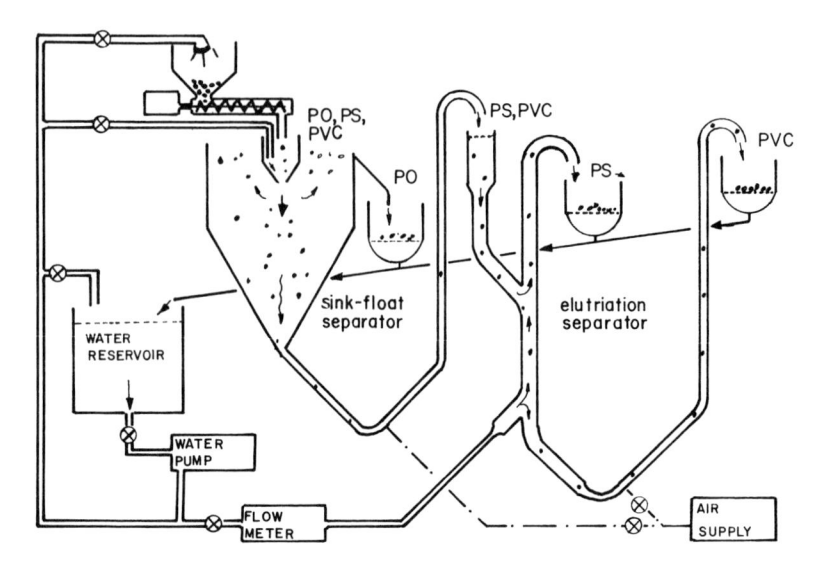

FIG. 3.22: Hydraulic separator. (Reprinted from Ref. 18, courtesy Bureau of Mines, U.S. Department of the Interior.)

compared to the PVC fraction. Most of the thermosets and composites remain with the heavy PVC fraction [18].

2. SEPARATION USING SELECTIVE WETTING CHARACTERISTICS

Although plastics are generally hydrophobic, their wetting characteristics can be selectively adjusted by the addition of surfactants.

TABLE 3.3 Analysis of Fractions Obtained from the Hydraulic Separation of Upgraded Plastic Concentrate from the Black-Clawson Process

Hydraulic separation		Sink-float analysis of fraction (wt % by density, g/cm^3)			
Fraction	(Wt %)	<1.0	1.0-1.2	1.2-1.5	>1.5
PO	48	97.0	3	--	--
PS	33	0.5	95	4.5	--
PVC	19	--	28	40.0	32

Figure 3.23 shows a typical set of contact angle curves. The contact angle decreases as the concentration of wetting agent increases. The effects of wetting agents are quite different on different plastics. The effect on PP is slight, but wetting increases on PE, PS, and PVC (in that order). The difference in wetting properties can be used as a basis of flotation separation. In mineral processing, flotation separation is achieved by air bubbles adhering to finely ground particles (50 to 100 µm). Because of the lower density of plastics, considerably larger particles can be floated. A flotation cell for plastics must satisfy the following conditions: (1) air bubbles of the most suitable size for the flotation of plastics must be uniformly generated; (2) agitation is necessary to avoid sedimentation of large particles; (3) there should be no turbulent water flow in the separation zone; and (4) the water surface must be smooth and stable.

Saitoh et al. [19] built a flotation cell meeting these requirements. Various kinds and shapes of plastics were mixed and fed into the cell. To effectively separate the mixtures, an appropriate wetting agent was added. After several minutes, an appropriate flotation promoter was added and air bubbles were

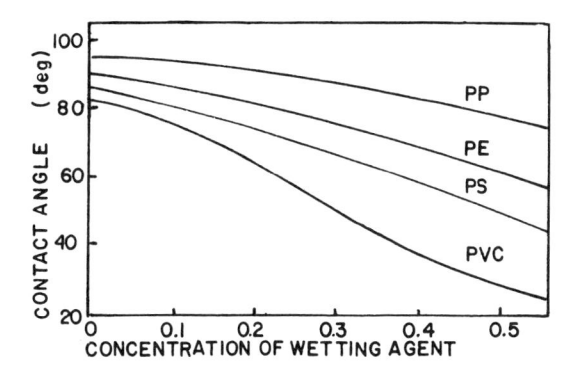

FIG. 3.23 An example of the relationship between concentration of wetting agent and contact angle. (Reprinted from Ref. 19, courtesy Elsevier Scientific Publishing Company.)

TABLE 3.4 Results of Batch Flotation Separation

Test samples	Composition (%)	Content of floated materials and recovery		Contents of residual matter and recovery	
PS pellets Rigid PVC granules	50 50	PS	98.4%	PVC Recovery rate	96.5% 98.4%
PP film Rigid PVC plate	50 50	PP	98.8%	PVC Recovery rate	97.8% 98.8%
MMA granules Rigid PVC granules	50 50	MMA	98.8%	PVC Recovery rate	97.3% 98.9%
PE film Flexible PVC sheet	50 50	PE	98.8%	PVC Recovery rate	97.3% 98.9%
NY pellets Rigid PVC granules	50 50	NY	98.5%	PVC Recovery rate	98.5% 98.5%
PS pellets PP film Flexible PVC sheet	34 33 33	PP	98.7%	PVC Recovery rate	97.4% 97.4%

Separation

PS pellets	20			PVC	95.9%
PP film	20	PP	98.9%		
PE film	20	PP			
Nylon film	20	PE		Recovery rate	95.9%
Rigid PVC granules	20				
PP granules	47	PP	97.0%	PS	99.0
PS granules	53	Recovery	98.9	Recovery rate	97.4%
PP granules	50	PP	98.8%	PE	97.3%
PE film	50	Recovery rate	97.3	Recovery rate	98.9
PET film	50	PET	100.0%	TAC	99.8%
TAC film	50	Recovery rate	97.0	Recovery rate	93.8%

Key: TAC = Cellulose triacetate; NMA = methyl methacrylate; NY = nylon; PET = polyester. Reprinted from Ref. 19, courtesy Elsevier Scientific Publishing Company.

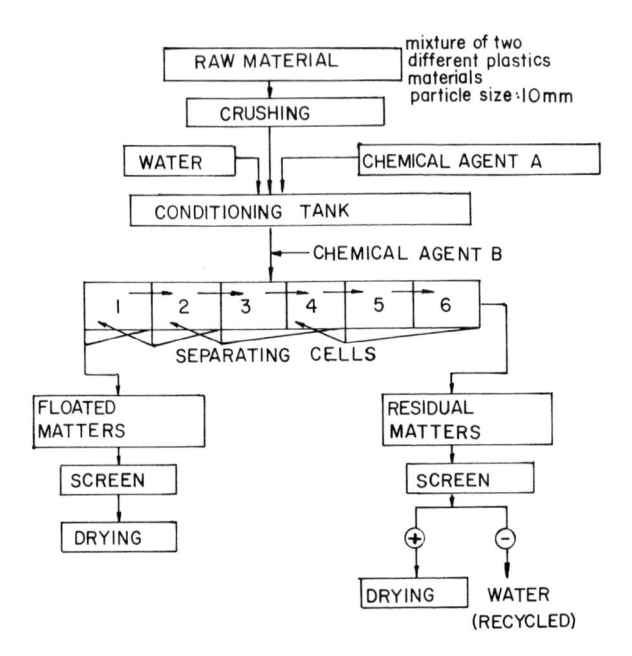

FIG. 3.24 Flowsheet of flotation separation process. (Reprinted from Ref. 19, courtesy Elsevier Scientific Publishing Company).

introduced into the cell. Some 2 to 10 min. were needed to achieve separation. The results of the batch separation given in Table 3.4 indicate a highly efficient separation.

A flowsheet of a continuous process for separating a mixture of two plastics is given in Fig. 3.24. Feed is ground into about 100 mm particle size. The feed, water, and surfactant are fed into the conditioning tank. After a few minutes the flotation promoter is added and the whole mixture is fed into separating cells where the separation takes place. Floating and sunken materials are removed, screened, and dried. Water is recycled back to the process [19].

3. *SOLVENT SEPARATION*

High-molecular-weight polymers are rarely soluble in one another. When two chemically different polymers are mixed they usually form

Separation

TABLE 3.5 Experimental Results of Solvent Separation

Run	Wt %					Total polymer (%)	% Cyclo-hexanone	Temp. (°C)	% PO in PO layer	% PS in PS layer	% PVC in PVC layer
	LDPE	HDPE	PP	PS	PVC						
1	45	21		17	17	17	15	125	98.8	96.9	98.4
2	45	21		17	17	10	15	115	99.5	98.6	98.9
3	36.3	16.5	13.2	17	17	10	15	125	80.8; 17.3[a]	97.1	97.9
4	45	21		17	17	10	10	125	98.5	96.8	97.4
5	45	21		12	22	10	15	125	97.8	97.7	98.8
6	66			17	17	10	15	125	99.0	97.5	98.3
7		66		17	17	10	20	120	98.9	98.8	98.3
8	41.6	19.4	19.5	19.5	10	10	15	125	99.1	99.1	98.8
9	45	21		17	17	12.5	15	125	99.5	98.9	99.1
10	45	21		17	17	12.5	20	115	99.3	99.2	99.5
11		66		17	17	12.5	20	125	99.2	98.5	98.7
12	41.6	19.4		19.5	17	19.5	12.5	115	99.5	99.5	99.3
13	45	21		17	17	15	15	125	99.9	99.5	99.6

TABLE 3.5 (continued)

Run	*	Wt %				Total polymer (%)	% Cyclo-hexanone	Temp. (°C)	% PO in PO layer	% PS in PS layer	% PVC in PVC layer
	LDPE	HDPE	PP	PS	PVC						
14	45	21		17	17	15	15	115	99.6	99.6	99.3
15	36.3	16.5	13.2	17	17	15	15	125	77.1; 22.4[a]	99.1	99.0
16	45	21		17	17	15	30	125	99.3	99.2	99.8
17	66			17	17	15	15	125	99.8	99.4	99.2
18	45	21		17	17	15	20	115	~100	99.8	99.3
19	41.6	19.4		19.5	19.5	15	20	115	99.5	99.6	99.9
20	45	21		22	12	15	20	125	99.2	99.2	99.4
21	48.4	22.6		14.5	14.5	15	20	115	99.8	99.7	99.5

[a]First number = %PE, second number = %PP.
Reprinted from Ref. 20, courtesy Society of Plastics Engineers.

separate phases. Even when a common solvent is used to dissolve each of the plastics, such solutions are not miscible. If a mixture of plastics is dissolved in a solvent, the several phases will contain solutions of almost pure components. Sperber and Rosen [20] studied the separation of a mixture of polyolefins (PO) PS, and PVC in various solvents. They found that a mixture of cyclohexanone and xylene makes an efficient solvent system. Table 3.5 shows the results of their experiments, which indicate that the major thermoplastics in a waste mix can be successfully separated. The effectiveness of the process is influenced by the following factors: composition of the solvents, temperature, type of waste feed, and the solvent/feed ratio [20].

REFERENCES

1. N. L. Drobny, H. E. Hull, and R. F. Testin, *Recovery and Utilization of Municipal Solid Waste; A Summary of Available Cost and Performance Characteristics of Unit Processes and Systems,* Report No. EPA-SW-10c-71, U.S. Environmental Protection Agency, 1971.
2. J. L. Pavoni, J. E. Heer, and D. J. Hagerty, *Handbook of Solid Waste Disposal,* Van Nostrand Reinhold, New York, 1975. pp. 294-345.
3. G. Shelef and M. Lapidot, "Recent Advances in the Treatment of Solid Wastes," IAEA, Munich, March 1975.
4. F. R. Jackson, *Recycling and Reclaiming of Municipal Solid Wastes,* Noyes Data Corp. 1975.
5. B. Citron and B. Halen, *Automated Recovery from Domestic Refuse,* Report STU 73-5182 U-4084-0002, AB Svenska Fläktfabriken.
6. J. F. Laundrie and J. H. Klungress, "Dry Methods of Separating Plastic Films from Waste Paper," *Paper Trade J.,* Feb. 5, 1973.
7. J. F. Laundrie, "Separation of Thermoplastic Film and Wastepaper," U.S. Pat. 3,814,240, 1972.
8. J. F. Laundrie and J. H. Klungress, "Effective Dry Methods of Separating Thermoplastic Films from Wastepaper," Research Paper FPL 200, U.S.D.A. Forest Service, 1973.
9. R. D. Fyfe, "Methods of Waste Thermoplastic Removal," U.S. Pat. 3,599,788.

10. Black-Clawson Fibreclaim, Inc., "Recovery of Plastics from Municipal Waste," U.S. Pat. 1,512,257, 1978.

11. M. R. Grubbs and K. H. Ivey, *Recovering Plastics from Urban Refuse by Electrodynamic Techniques,* Technical Progress Report 63, U.S. 1972.

12. C. D. Belcher, "Reclaiming Fiber Supported PVC Scrap," *SPE J. 29,* June 1973.

13. W. B. Sussman, C. D. Belcher, and G. E. Brown, "Method for Reclaiming Commefcially Useful Fibers and Resin from Scrap Material," U.S. Pat. 3,624,009, 1971.

14. E. A. Hafner, "Vinyl Chloride Polymer Recovery Process," U.S. Pat. 3,836,486, 1974.

15. E. Wainer, "Recovery of Flexible and Rigid Materials from Scrap Polyvinylchloride, Its Copolymers and Cogeners," U.S. Pat. 3,912,664, 1975.

16. A. J. Felton, "The Process and Economics of Polymer-coated Woods Fibre Recovery," *Tappi 58(5),* 1975.

17. C. S. Marty, "Contaminant Elimination Through Solvent Extraction," Pulp and Paper Seminar, *Fiber Conservation Utilization Proceedings,* Chicago, Miller Freeman Publications, San Francisco, May 1974.

18. J. L. Holman, J. B. Stevenson, and J. Adam, Recycling of Plastics from Urban and Industrial Refuse, Report of Investigations 7955, U.S. Bureau of Mines.

19. K. Saitoh, J. Nagano, and S. Izumi, "New Separation Technique for Waste Plastics," *Res. Rec. Cons. 2,* 1976.

20. R. J. Sperber and S. L. Rosen, "Recycling of Thermoplastic Waste: Phase Equilibrium in Polystyrene – PVC – Polyolefin Solvent Systems," *Polym. Eng. Sci. 16(4),* 1976.

Chapter 4

PRIMARY RECYCLING

I. DEGRADATION OF THERMOPLASTICS DUE TO REPETITIVE PROCESSING

Primary recycling involves using uniform, uncontaminated plastic waste to manufacture plastic products. Only thermoplastic waste can be directly reprocessed: it can be used alone or, more often, added to virgin resin at various ratios. Primary recycling can be performed by the processor in plant or through outside reprocessors. The main technical problems encountered in primary recycling are (1) degradation of the material due to repeated processing, resulting in a loss in such properties as appearance, chemical resistance, processability, and mechanical characteristics; (2) contamination of the reprocessed plastic; and (3) handling of low-bulk-density scrap such as film or foam.

A. Mechanisms

Changes in the physical properties of plastics, observed after processing at elevated temperatures, are almost entirely due to changes in polymer structure. Changes that may occur in the molecular structure are: (1) reduction of the average molecular weight, (2) increase of the average molecular weight due to cross-linking, and (3) formation of unsaturation or cyclization due to side-chain reactions. Thermal and thermo-oxidative degradation and cross-linking are the most important factors involved. A decrease in the molecular weight in the case of main chain scission can occur randomly or at specific sites. Rupture at the ends of molecular chains (resulting in only small change in molecular weight) has only a slight effect on physical properties. Evolution of volatile components, however, might cause secondary reactions as well as processing problems. Radical combination during thermal degradation may lead to branched structures which can eventually cross-link.

Reactions including side chains and substituents may lead to cyclization or unsaturation, both resulting in stiffening of the molecular chain [1]. A typical example of the thermal degradation reaction involving a side chain is dehydrohalogenation of PVC. At elevated temperature hydrogen chloride may be removed, leaving an unsaturated structure. The hydrogen chloride then catalyzes further degradation.

Oxygen usually cannot be excluded from the polymer melt during processing, resulting in thermo-oxidative degradation. Thermal oxidation may lead to cross-linking (resulting in hardening of the polymer) or to chain scission (resulting in a softer polymer). The most important feature of that process is the free radical reaction, leading to the incorporation of many molecules of oxygen from a single free radical:

$$R\cdot + O_2 \rightarrow ROO\cdot \quad (1)$$

$$ROO\cdot + RH \rightarrow ROOH + R\cdot \quad (2)$$

Hydroperoxides formed in this reaction are unstable at processing temperatures. The decomposition of hydroperoxides forms new free radicals:

$$2ROOH \rightarrow ROO\cdot + H_2O \quad (3)$$

This gives rise to the autoaccelerating rate curve, which is eventually balanced by the termination reaction.

$$2ROO\cdot \rightarrow \text{inert products} \quad (4)$$

Commercial plastics contain stabilizers such as hindered phenols or aromatic amines which trap or remove free radicals:

$$ROO\cdot + AH \rightarrow ROOH + A\cdot \quad (5)$$
$$2A\cdot \rightarrow \text{inert products} \quad (6)$$

The presence of antioxidants extends the induction period of the autoaccelerating thermo-oxidative process [2].

Reinforced thermoplastics rely on fibrous or platelike fillers for a considerable portion of their mechanical properties. Degradation of the mechanical properties of reinforced plastic during processing is due to three factors: (1) degradation of the polymer, (2) degradation of the polymer/reinforcement interface, and (3) breakdown of the reinforcement. Degradation of the polymer is basically similar to that of the unreinforced material, although in a few cases a specific reinforcing agent or filler might accelerate the degradation reaction. Nothing is known about the effect processing has on the properties of the interface. In most cases the effect is probably minimal.

The effectiveness of the reinforcement depends on the aspect ratio (the ratio of length to diameter) of the reinforcing agent. High-shear processing (such as extrusion or injection molding) causes breakage of the reinforcement and a decrease in aspect ratio. During prolonged processing at a given shear rate, the average fiber length decreases exponentially to an asymptotic limit; further processing does not result in further fiber degradation. The value of this limiting aspect ratio depends on the mechanical properties of the reinforcement and the shear rate and stress of the processing. Breakage of the reinforcement results in lower mechanical properties.

Degradation of plastics during elevated temperature processing can be manifested in a number of ways, such as:

Change of the melt viscosity (increase due to cross-linking or decrease due to molecular chain breakage)
Change of the physical properties such as strength, impact resistance, stiffness
Change in color
Reduction of chemical resistance

Not all plastics are equally sensitive to degradation during processing. Oxidation is the mode of breakdown of high- and low-density PE, manifested in reduction of melt flow. Recycling of PP results in an increase in melt flow and a falloff in impact strength. PP is also sensitive to contamination, promoting degradation signified by an increase in brittleness, dark streaks, burn marks, and odor. PS is also sensitive to contamination during reprocessing, causing a reduction in some properties and some color shift.

On prolonged recycling of PVC the stabilizers become depleted and the material begins yellowing and emitting odors. Reprocessing of acrylics causes some color shift; however, for less critical applications, up to 100% reprocessed material may be used. Nylons can be easily reprocessed, although they are sensitive to contamination and tend to "brown" with repeated reprocessing. The properties of nylons are usually maintained over a wide range of processing conditions.

Clean reground ABS can be mixed with virgin material in any proportion. The ratio of virgin to reground material, however, should be kept constant through the production run, otherwise the melt temperature and production rate may vary. Degradation of reworked ABS is manifested by a falloff in physical properties, particularly impact strength and darkening of the material. Acetal resins are sensitive to contamination and degrade when subjected to prolonged processing. The homopolymer is more sensitive than the copolymers. Usually up to 15% regrind is used [3].

Figure 4.1 shows the generalized property changes resulting from different degradation mechanisms during processing. An example of linear decrease in properties (curve 1) is the notched impact strength of polycarbonate. The notched impact strength of glass reinforced nylon-6/6 follows curve 2. Curve 3 is typical of the properties change of plastics containing thermal stabilizers. If the degradation process involved the formation of products that promote further degradation, the rate of degradation increases with the number of recycles (curve 4) [1]. Yen [4] gives examples

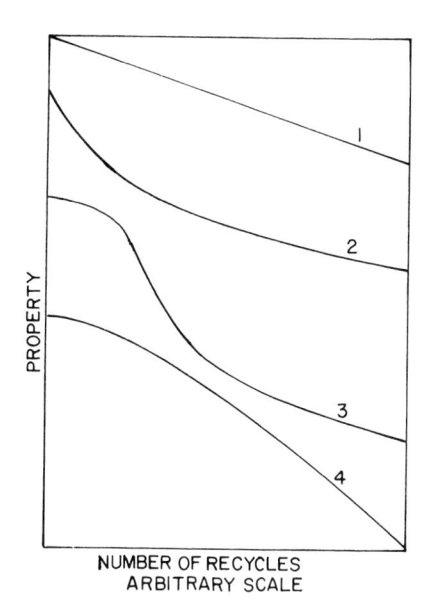

FIG. 4.1 Property changes as a result of different degradation mechanisms during processing. (Reprinted with permission from K. B. Abbas, A. B. Knutsson, and S. H. Berglund, "New Thermoplastics from Old," Chemtech, Aug. 1978. Copyright by the American Chemical Society.)

of the effect on properties of recycling some plastics (Figs. 4.2, 4.3, and 4.4). The curves represent only one recycling cycle. If the process is continued, further changes in properties will occur. The figures show that various materials, or various properties of the same material, react differently to processing, the type of process and processing conditions also influence the properties of parts containing regrind.

B. Mathematical Models Predicting Changes in Properties

A flowsheet of an in-plant primary recycling process is shown in Fig. 4.5. Converting equipment converts feed material consisting of virgin and recycled plastics into a product. Some waste is also produced, most of which is recycled; a certain proportion of the waste may have to be discarded. After one cycle the product stream

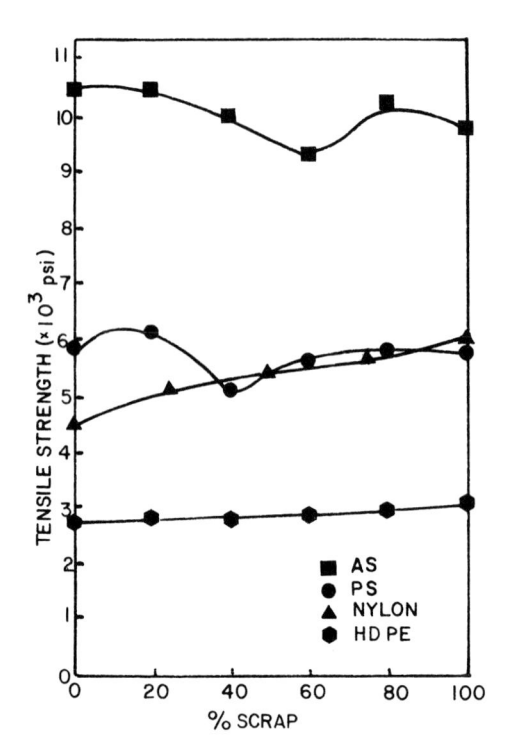

FIG. 4.2 Tensile strength as a function of percent regrind. (Reprinted from Ref. 4, courtesy of Society of Plastics Engineers, Inc.)

consists of the material that has undergone one and the material that has undergone two processing cycles. The composition of the product stream after one cycle is:

$$ps = F(k + r) \qquad (7)$$

where ps is product stream in lb/hr (both product and waste, for example, molded parts and sprues and runners), F is the feed rate in lb/hr (both virgin and recycled), k is the ratio of virgin to total feed, and r is the ratio of recycled material to total feed. After the second cycle, the product stream is

$$ps = F[k + r(k + r)] = F(k + kr + r^2) \qquad (8)$$

Primary Recycling 125

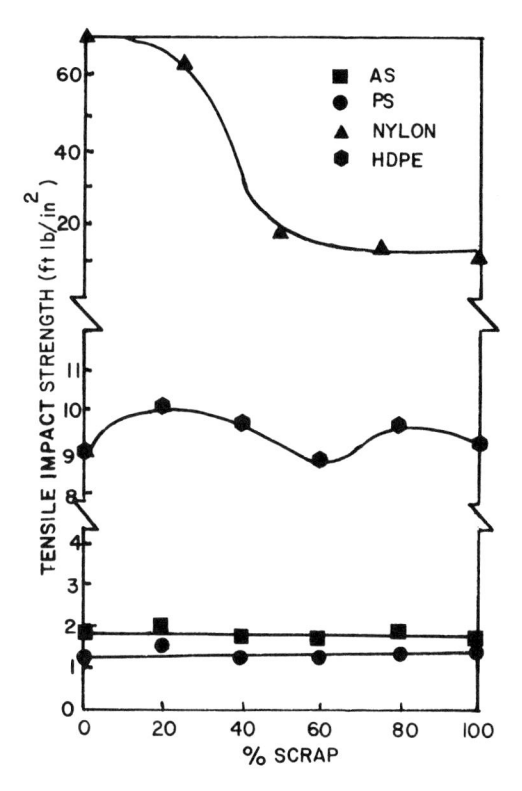

FIG. 4.3 Tensile impact strength as a function of percent regrind.
(Reprinted from Ref. 4, courtesy of Society of Plastics
Engineers, Inc.)

where Fk represents the portion of the material which has been
subjected to one processing cycle, Fkr to two cycles, Fr^2 to three
cycles. After n cycles, the composition of the products stream is
given by

$$ps = F \left(\sum_{j=0}^{n-1} kr^j + r^n \right) \qquad (9)$$

Figures 4.6, 4.7, and 4.8 show the percentage of original material
present in the product after various numbers of recycling cycles
with 50, 25, 10, and 5% recycle, 10% recycle and waste, and 50%

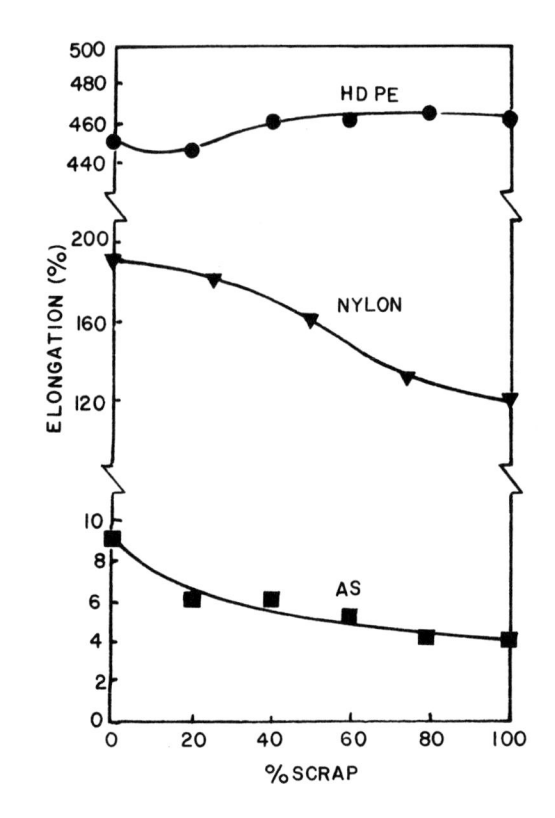

FIG. 4.4 Elongation as a function of percent regrind. (Reprinted from Ref. 4, courtesy Society of Plastics Engineers, Inc.)

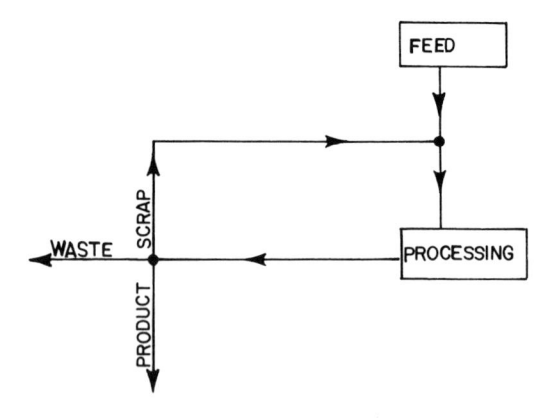

FIG. 4.5 Diagram of recycling process.

Primary Recycling

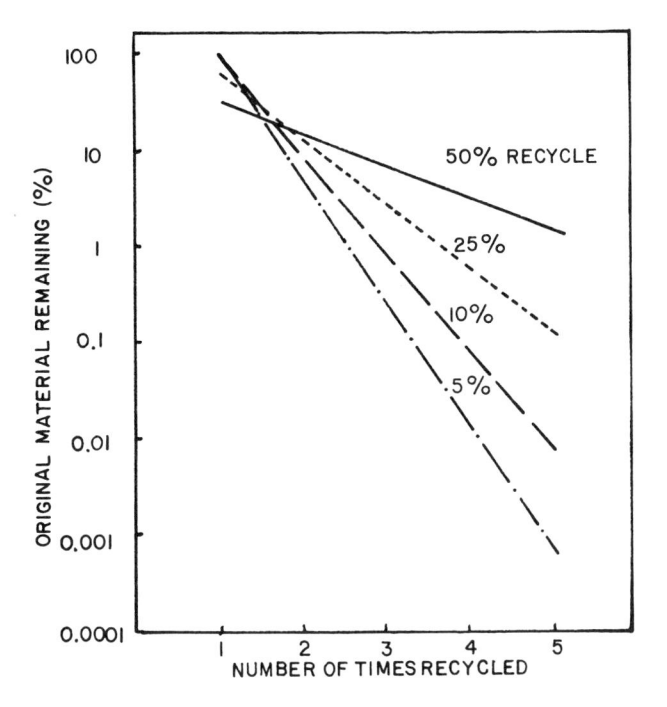

FIG. 4.6 Depletion of original material after a given number of recyclings. (Reprinted from Ref. 5, courtesy Society of Plastics Engineers, Inc.)

recycle and waste. These graphs are based on Eq. (9). On a semi-log graph these plots form straight lines. Figure 4.9 shows the percent of original material present as a function of the percent of recycled material used in the feed. At 10% recycled material used, after four cycles there is only 0.01% of the original material (that is material which went through all four processing cycles) present. If, however, the amount of recycled material is increased to 50%, the amount of original material present in the part will increase to almost 10% [5].

The presence of a large amount of material with a long heat history will affect the properties of the product. Abbas et al. [1] developed a mathematical model for calculating the change in

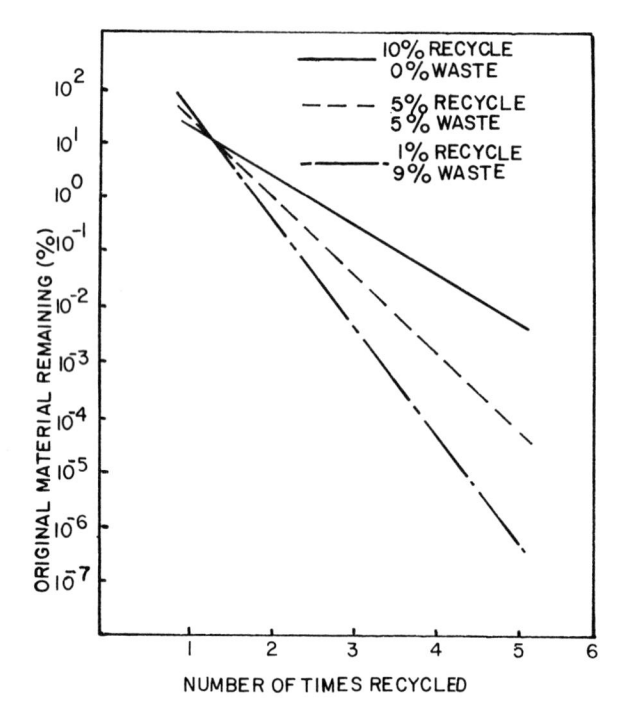

FIG. 4.7 Depletion of original material after a given number of recyclings. (Reprinted from Ref. 5, courtesy Society of Plastics Engineers, Inc.)

properties as a function of the number of cycles and the percentage of regrind. The calculations were performed for two practical cases:

1. Scrap is ground and stored separately. A certain percentage of virgin material is always added before the regrind is processed.
2. Scrap is ground directly at the injection molding machine and collected in a bag. Intermittently, the content of the bag is mixed with the virgin material and reprocessed.

Let us consider the latter case first. The property of the material collected in the first bag is P_1. The property of the

Primary Recycling

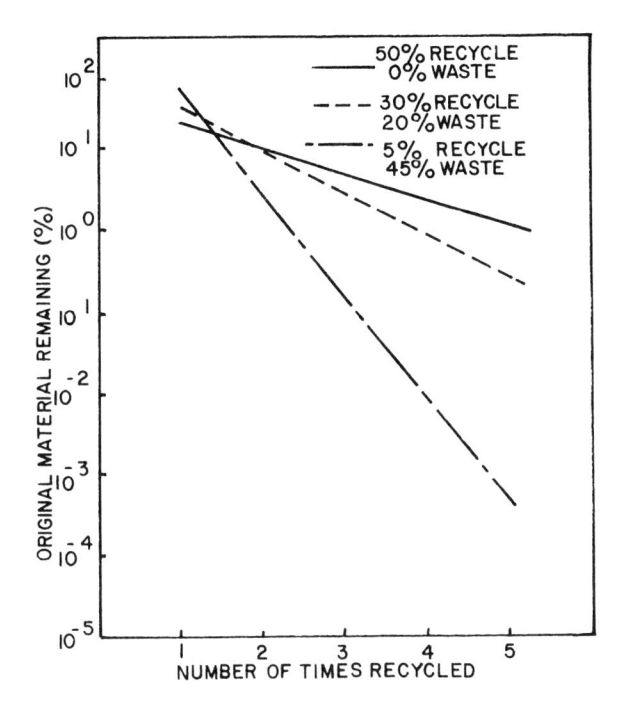

FIG. 4.8 Depletion of original material versus percent recycled material used in the feed. (Reprinted from Ref. 5, courtesy of Society of Plastics Engineers, Inc.)

material collected in the second bag will be

$$P_1 c + P_2 (1 - c) \tag{10}$$

where c is the ratio between the weight of a salable product (part weight) and shot weight, and P_2 is a property of a given plastic after two processing cycles. The properties of the material in the nth bag will be

$$P_1 c + P_2 c(1 - c) + P_3 c(1 - c^2) + \cdots + P_n (1 - c)^{n-1} \tag{11}$$

When the nth bag is mixed with a fraction k of virgin material (P_o) in the proportion $(1 - k)/k$ the property at steady state P_s will be

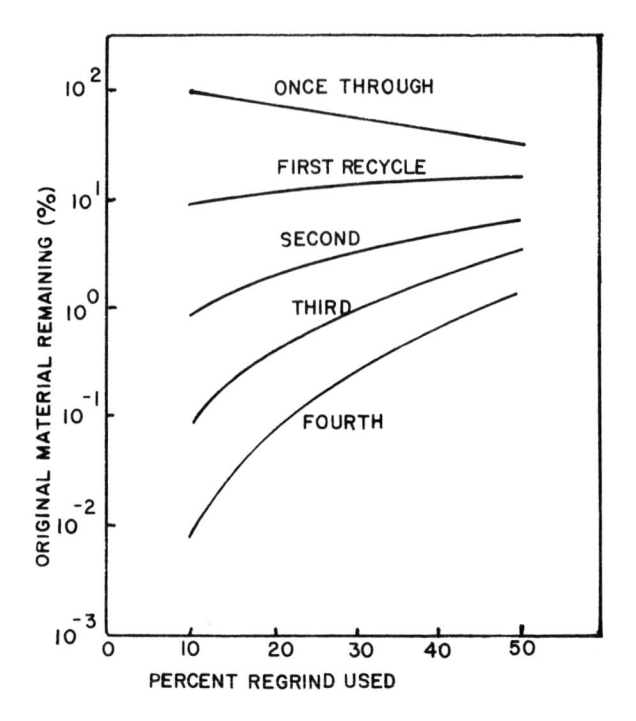

FIG. 4.9 Depletion of original material versus percent recycled material used in the feed. (Reprinted from Ref. 5, courtesy of Society of Plastics Engineers, Inc.)

$$\begin{aligned} P_s &= kP_0 + (1-k)[cP_1 + c(1-c)P_2 + c(1-c)^2 P_3 + \cdots \\ &\quad + c(1-c)^{n-2} P_{n-1} + (1-c)^{n-1} P_n] \\ &= kP_0 + (1-k)c \sum_{j=1}^{n-1} (1-c)^{j-1} P_j \\ &\quad + (1-k)(1-c)^{n-1} P_n \end{aligned} \qquad (13)$$

If a constant fraction of virgin material is always added to the regrind before reprocessing, the steady state can be calculated from

Primary Recycling 131

$$P_s = kP_o + k(1-k) \sum_{j=1}^{n-1} (1-k)^{j-1} P_j + (1-k)^n P_n \quad (14)$$

If the property change due to processing of a given material is known, the property change in the case of various numbers of recycling cycles or recycling ratios can be computed from Eq. (13) or (14).

Figure 4.10 shows a typical example of the change in properties as a function of the number of processing cycles. Two distinct degradation behaviors are shown: linear and nonlinear. Curves like those in Fig. 4.10 can be obtained by processing the material, testing its properties, grinding the samples, processing them again, and so on. Values from Fig. 4.10 were substituted in Eq. (14) (for nonlinear behavior a logarithmic fit was used) to compute the properties of samples after a steady state had been achieved. The results are shown in Fig. 4.11. To retain 90% of the original property, only 20% of the regrind can be added to the "nonlinear" material with the higher initial degradation rate (curve 3), but

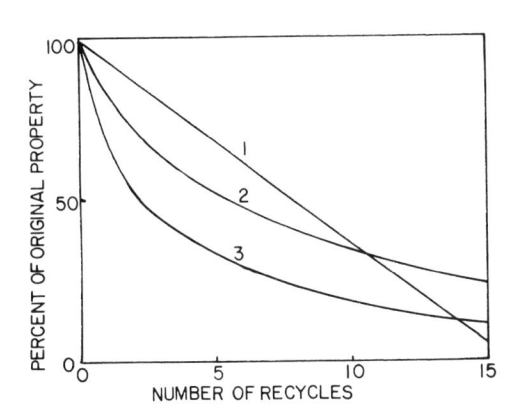

FIG. 4.10 Property decay curves used for comparison of linear and nonlinear relationships. (Reprinted with permission from K. B. Abbas, A. B. Knutsson, and S. H. Berglund, "New Thermoplastics from Old," *Chemtech*, Aug. 1978. Copyright by the American Chemical Society.)

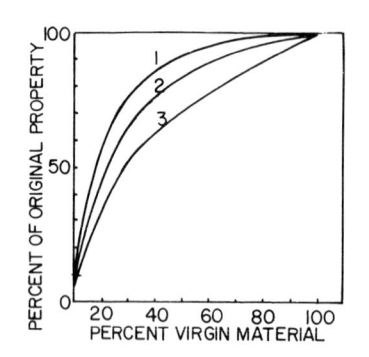

FIG. 4.11 Percent original property as a function at percent virgin material. Data from Fig. 4.10 have been used in calculations. (Reprinted with permission from K. B. Abbas, A. B. Knutsson, and S. H. Berglund, "New Thermoplastics from Old," *Chemtech*, Aug. 1978. Copyright by the American Chemical Society.)

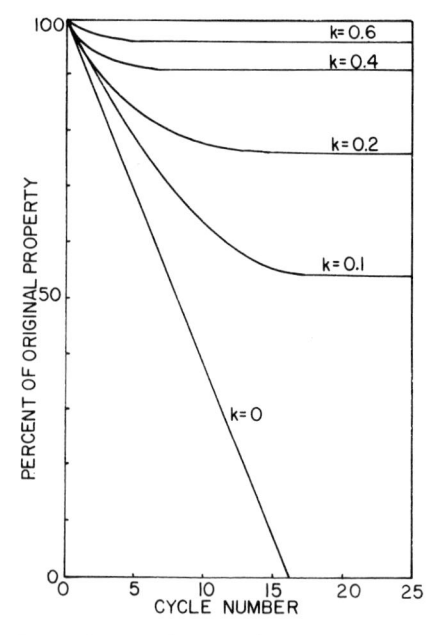

FIG. 4.12 Changes in properties as a function of cycle number and fraction of virgin material. (Reprinted with permission from K. B. Abbas, A. B. Knutsson, and S. H. Berglund, "New Thermoplastics from Old," *Chemtech*, Aug. 1978. Copyright by the American Chemical Society.)

up to 60% of the regrind of the "linear" material will give the same 90% property retention (curve 1). Figure 4.12 shows the dependence of the property decrease on k (fraction of virgin material added to the regrind). The straight line represents 100% regrind and is identical to the straight line in Fig. 4.10. The addition of only 10% virgin material (k = 0.1) increases the steady state to 54%. A 95% property level can be maintained at k = 0.6 and a 90% property level at k = 0.4 [1].

The tensile strength of a thermoplastic reinforced with short aligned fibers can be calculated from the Kelly-Tyson equations (6):

$$\sigma_{uc} = \sigma_f V_f \left(1 - \frac{L_c}{2L}\right) + \bar{\sigma}_m V_m \qquad L \geq L_c \qquad (15)$$

$$\sigma_{uc} = \frac{\tau L}{d} V_f + V_m \sigma_m \qquad L < L_c \qquad (16)$$

where σ_f is the tensile strength of the fiber, V_f is the volume fraction of the fibers, $\bar{\sigma}_m$ is the stress in the polymer strained to the ultimate strain of the fibers, V_m is the volume fraction of the matrix, σ_m is the ultimate stress of the polymer, τ is the yield shear strength of the interface, L is the length of the fibers, d is the diameter of the fibers, and L_c is the critical length of the fibers (if $L < L_c$, when the composite breaks the fibers pull out from the resin, but do not break; if $L \geq L_c$, the composite fails when the fibers fail). Let us assume that the average fiber length after the Nth 100% regrind molding $\overline{L(N)}$ can be expressed as

$$\overline{L(N)} = a \exp(-\alpha N) + b = \{\overline{L(N-1)} - b\} \exp(-\alpha) + b \qquad (17)$$

where $a + b = \overline{L(0)}$ is the initial average fiber length and b is the minimum fiber length to which the fiber can be degraded under the given shear stress and shear rate conditions. If the ratio of regrind to total feed is designated r, the average fiber length before Nth molding $\overline{L(N-1)}'$ can be expressed as

$$\overline{L(N-1)}' = \overline{L(N-1)}r + \overline{L(0)}(1-r) \qquad (18)$$

By combining equations (17) and (18), the average fiber length after Nth processing can be calculated:

$$\overline{L(N)}' = \{\overline{L(N-1)}' - b\}e^{-\alpha} + b$$
$$= \{\overline{L(N-1)}r + \overline{L(0)}(-r) - b\}e^{-\alpha} + b$$
$$= ae^{-N\alpha} + a(1-r)e^{-\alpha} + b \qquad (19)$$

For processing with regrind ratio r and assuming that the change of the properties of the matrix due to processing is minimal, the composite tensile strength can be estimated from

$$\overline{\sigma_c(N)}' = \eta\sigma_f V_f \left(1 - \frac{L_c}{2\overline{L(N)}}\right) + \overline{\sigma_m} V_m \qquad \overline{L(N)}' \geq L_c \qquad (20)$$

$$\overline{\sigma_c(N)}' = \eta\frac{\tau\overline{L(N)}'}{d}V_f + V_v \sigma_m \qquad \overline{L(N)}' < L_c \qquad (21)$$

where η is a correction factor accounting for the three-dimensional distribution of the fibers. Figure 4.13 shows the dependence of the ultimate strength of a typical fiber-reinforced thermoplastic (nylon-6/6, 20 wt % of glass fibers) on the number of recycles and regrind ratios. When r = 0, no regrind is used and the tensile strength is independent of the number of moldings. As r increases, the dependence becomes more evident and the tensile strength at steady state decreases sharply with increasing amount of regrind material [7].

The mathematical models just presented are useful in predicting the property losses in primary recycling, in establishing the percentage of regrind allowable, or the necessary degree of polymer stabilization.

II. INDUSTRIAL PRACTICE

A. Granulators

In order to be reprocessed, plastics waste has to be ground to a particle size close to that of virgin resin. The size reduction

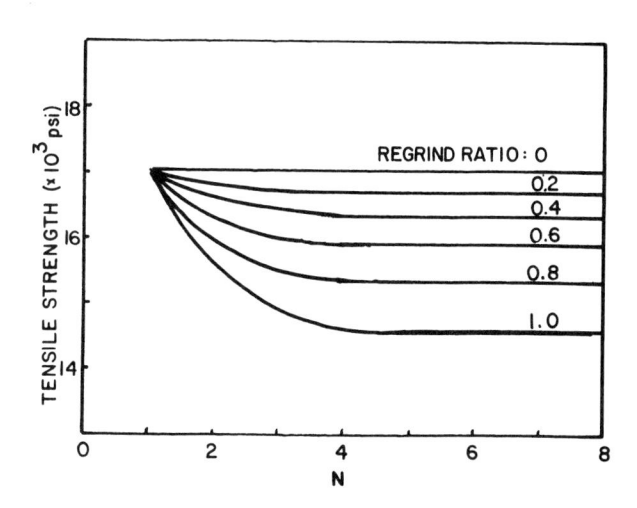

FIG. 4.13 Dependence of the ultimate tensile strength at a typical fiber reinforced thermoplastic on the number of recycles and regrind ratios. $L(0) = 30.2$ mil, $L_c = 8$ mil, $F_f = 0.11$, $a = 22$ mil, $b = 8.2$ mil, $d = 1.35$, $\eta \sigma_f V_f = 1.15 \times 10^4$ psi, $\overline{\sigma}_m V_m = 8.9 \times 10^3$ psi. (Reprinted with permission from Ref. 7. Copyright 1979 by John Wiley & Sons, Inc.)

in most cases is accomplished with the use of a granulator. Various types of granulators are used, depending on the required throughput, the size of the pieces to be granulated, and the physical shape of the plastic waste (film, pipe, sheet, sprues). Each granulator consists of a hopper, cutting chamber (rotor with the knives), screen, and drive. The shape of the hopper is designed to handle specific shapes. Granulators may be equipped with a feeding device, such as an auger feeder or counterrotating rollers for film. Most hoppers have bends, baffles, and/or doors to prevent fly-back of the plastic pieces.

Granulation takes place in the cutting chamber, which is enclosed with very strong walls: on some larger models, up to 4-in. thick steel plates are used. The geometry of the feed entrance is often emphasized by the manufacturers' possible geometrics are shown on Fig. 4.14. In the "straight down" arrangement the waste

"STRAIGHT DOWN"

TANGENTIAL FEEDING

VERTICAL ROTOR

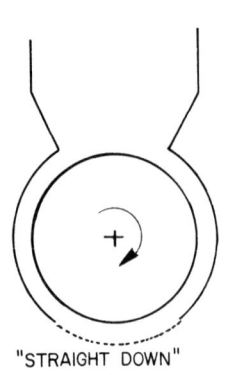

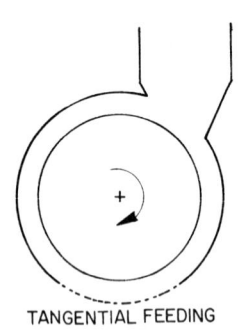

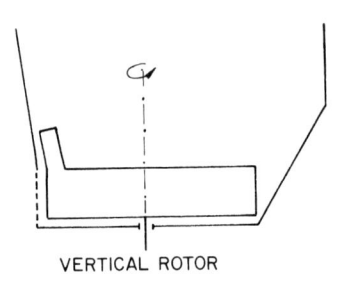

FIG. 4.14 Feed entrance geometrics.

Primary Recycling 137

is caught by the downstroke of the rotor and cut; a disadvantage is that it can be struck by the upstroke and thrown back into the hopper. This problem is minimized by offsetting the feed entrance toward the downstroke. In the vertical rotor design the rotor axis is vertical and the parts are fed from the top in such a way as to be exposed to the entire cutting diameter. A rotating tear knife helps pull material into the chamber.

Figure 4.15 shows two ways of mounting the knives: radial and tangential. Tangential mounting is generally used for light-duty cutting, especially for softer materials in which sheer cut reduces heat generation. Radial mounting results in a breaking rather than a cutting action. Figure 4.16 shows the most common knife designs. The knives are mounted on open rotors comprised of a series of rib-like supports, or on solid rotors. Figure 4.17 shows rotor designs with increasing numbers of knives. More knives increase throughput but also decrease the effective "bite size" of the granulator. Two- and three-blade rotors are most common.

The rotating knives cut against the stationary bed knives. Two bed knives are commonly used, but a greater or lesser number is also possible (Fig. 4.18).

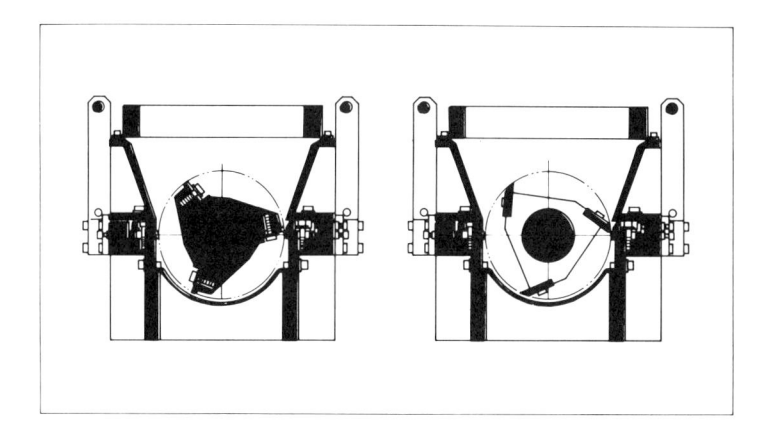

FIG. 4.15 Two basic designs of knife mounting; radial mounting and solid rotor (1) and tangential mounting with open rotor (r). (Reprinted from Ref. 8, courtesy Hartman Communications, Inc.)

138   Plastics Waste

RADIAL ROTOR KNIFE

STEEP-ANGLE GRANULATOR KNIFE

keen-edge rotor knife

reverse-bevel rotor knife

HOOKED ROTOR KNIFE

FIG. 4.16 Common knife designs. (Reprinted from Ref. 8, courtesy Hartman Communications, Inc.)

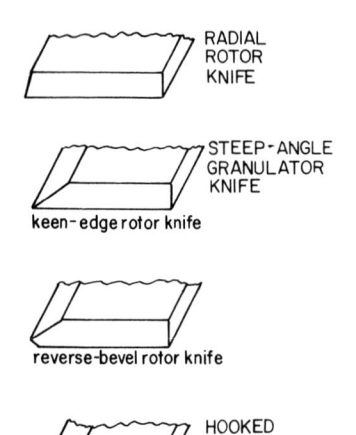

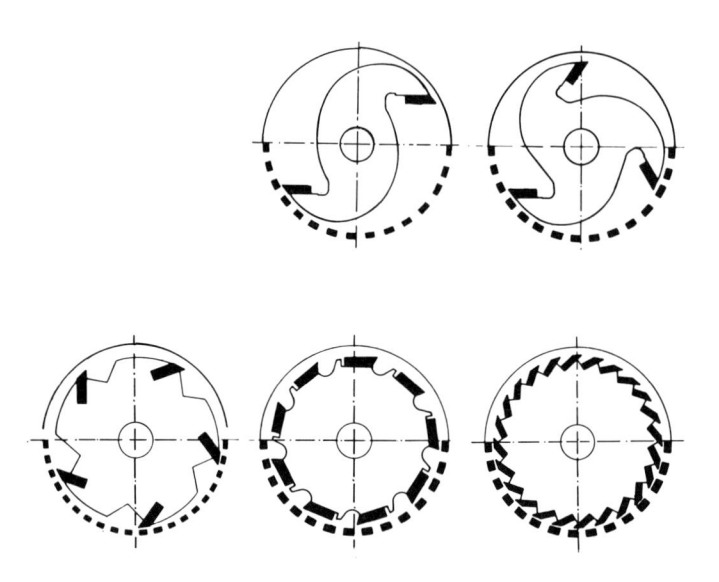

FIG. 4.17 Rotor design with an increasing number of knives. (Reprinted from Ref. 9, courtesy Pallman Machinenfabrik, GmbH & Co. KG.)

Primary Recycling   139

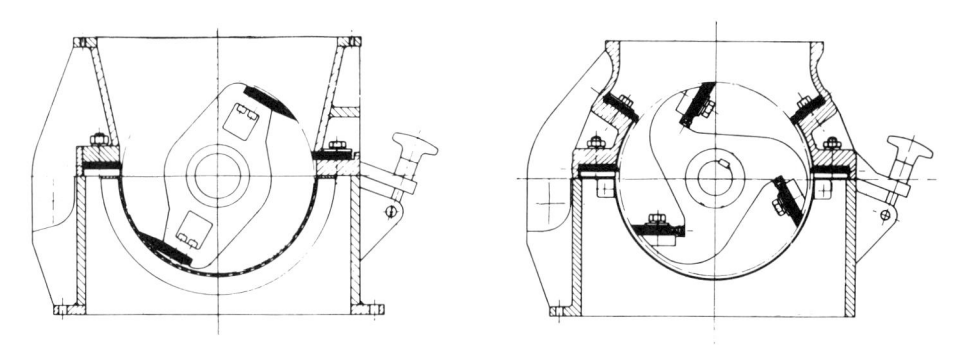

FIG. 4.18   Knife mills with two and four bed knives. (Reprinted from Ref. 9, courtesy Pallman Machinenfabrik, GmbH & Co. KG.)

PARALLEL KNIVES

SLANTED ROTATING KNIFE

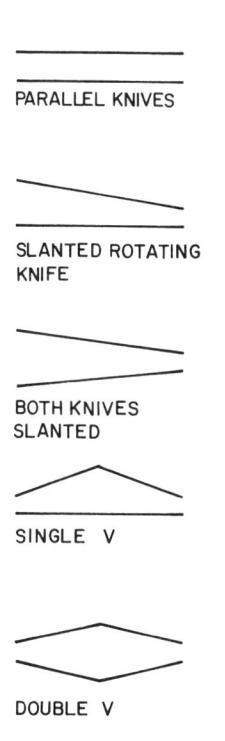

BOTH KNIVES SLANTED

SINGLE V

DOUBLE V

FIG. 4.19   Various knife arrangements.

FIG. 4.20   Various knife arrangements.

   The position of the blade is a very important design parameter: various possible arrangements are shown on Fig. 4.19.
   "Scissor-cut" granulators are increasing in popularity. Scissor action is accomplished by slanted rotating knives. Granulation of large waste pieces can be accomplished by designing the rotor so that it gradually removes the small pieces. This can be done by using rotors with staggered knives (Fig. 4.20). Other designs are also possible. Figure 4.21 shows a grinder with two counterrotating rotors; no stationary knives are used in this design. Figure 4.22 shows the feeding section of a two-stage design. The feeding section breaks large pieces which are fed into the cutting chamber for final grinding. Output of the granulator is influenced by screen area and opening size, which may range from 1/4 to 3/8 in. in diameter. Plugging may be reduced by chambering the holes and positioning an air jet below the bed knife.

FIG. 4.21   Two counterrotating granulator rotors.

FIG. 4.22   Feed section of a two-state granulator.

141

The rotor may be driven directly from the motor or by means of a V belt. Direct drives tend to be the most rugged, while the belt provides a measure of safety by protecting the motor from cutting shocks. Granulators are equipped with safety devices that prevent access to the cutting chamber while the rotor is in motion. Since the rotor continues to spin from inertia after being shut off, some granulator manufacturers fasten the hopper with such a long bolt that the rotor is dead by the time the hopper is opened. In the larger, hydraulically opened granulators, a delay timer may be used to actuate the hydraulics after the rotor has stopped [8].

B. Granulating "Difficult" Materials

Some plastics wastes are difficult to grind with standard granulators because of their physical properties (e.g., rubbery or low-melting-point materials) or because of their low bulk density (e.g., foams, films, fibers). Such materials must be granulated using special techniques, the most important of which are described below.

1. CRYOGENIC GRINDING

In cryogenic grinding the materials to be ground are embrittled by cooling to very low temperatures. The technique is used for plastics waste where finely powdered regrind of -30 mesh or finer is required, and when the plastic to be ground is tough and tends to generate enough heat to melt during grinding at normal temperatures. The three means of cooling materials before grinding are with liquid nitrogen, with liquid and solid carbon dioxide, and by mechanical refrigeration. Because of the relatively slow heat transfer, mechanical refrigeration is not commercially practical. Liquid and dry CO_2 have been used, but have several limitations compared with liquid nitrogen. Liquid nitrogen (LIN) systems can be controlled much better than CO_2 systems resulting in improved heat transfer efficiency and lower total grinding costs.

Primary Recycling 143

Liquid nitrogen can be applied by direct injection into the grinding chamber, by precooling the feed in a feed hopper, and by precooling the material on a special conveyor. Using a cooling conveyor usually results in maximum refrigeration efficiency. Liquid nitrogen at -320°F is sprayed directly on the plastic to be ground and quickly evaporates, absorbing heat from the plastic. Additional nitrogen may be injected into the grinding chamber to remove heat resulting from the grinding operation. Systems have been developed by the Union Carbide Corporation (Fig. 4.23) and by Air Products and Chemicals Inc. (Fig. 4.24). In Union Carbide's system, liquid nitrogen is injected into a precooler at a point close to the discharge to the mill. Nitrogen vapor travels against the movement of the material, paddles on the precooler shaft facilitating the mixing of material and coolant. Liquid nitrogen is also injected into the grinder. The Air Products system uses both co-

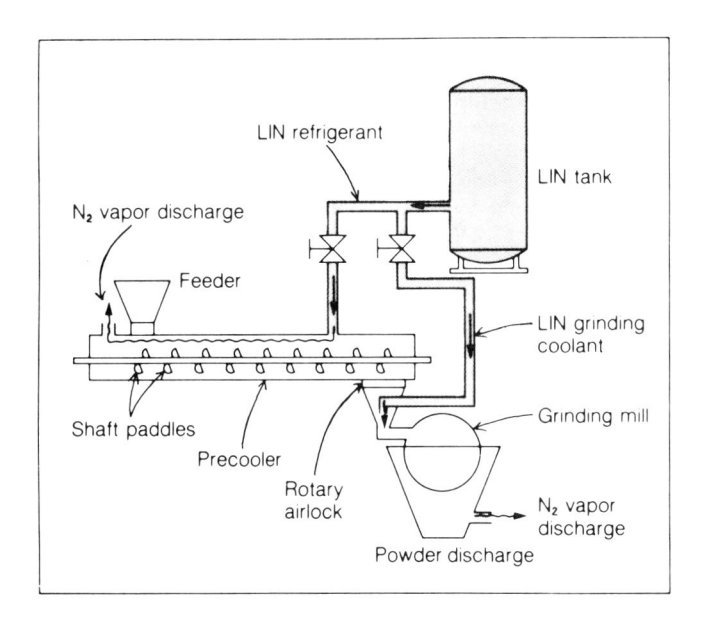

FIG. 4.23 Union Carbide's cryogenic grinding system. (Reprinted from Ref. 10, courtesy McGraw-Hill, Inc.)

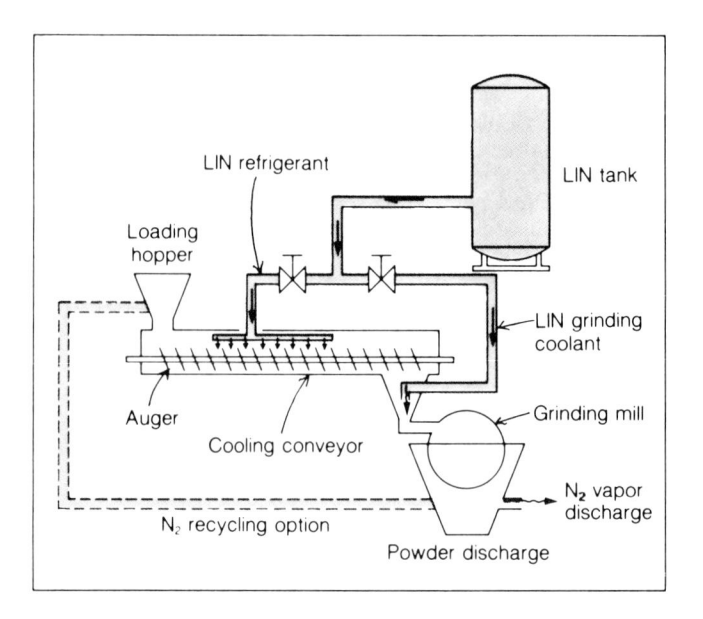

FIG. 4.24 Air Products' cryogenic grinding system. (Reprinted from Ref. 10, courtesy McGraw-Hill, Inc.)

TABLE 4.1 Cryogenic Grinding Data for Various Waste Plastics

Plastic material	Grinding temp. (°F)	LIN consumed (lb/lb product)	Production rate (lb/hp-hr)	Mesh size
Rigid PVC	−50 to +50	0.6 to 1.2	15 to 45	−20 to −50
Flexible PVC	−60 to 0	0.9 to 1.6	25 to 30	−30 to −60
Agglomerated PVC	+50 to +75	0.2 to 0.5	25 to 30	−40 to −50
Polyethylene (film, etc.)	−180 to −80	1.2 to 2.9	9 to 18	−20 to −40
Polyurethane	−200 to −30	1.5 to 4.0	9 to 31	−30

Reprinted from Ref. 11, courtesy Society of Plastics Engineers, Inc.

Primary Recycling 145

and countercurrent flows of nitrogen, while a specially designed conveyor screw facilitates mixing. Nitrogen from the cooling conveyor gets into the mill and helps dissipate the heat of grinding. Table 4.1 gives typical cryogenic grinding data for various waste plastics. On the average, the process requires one pound of LIN for one pound of material processed [10,11].

2. REPROCESSING LOW-BULK-DENSITY PLASTICS WASTE

Low-bulk-density plastics waste such as film, fiber, or foam may be reprocessed by grinding, followed by extrusion using a specially designed screw or compactors as a part of a feeder, or by pelletizing by using the heat of grinding to melt and agglomerate the particles. Standard processing equipment is designed to process plastic pellets or, less often, powders, and low-bulk-density materials cannot be efficiently processed on such equipment. The problem is overcome by the use of stuffing hoppers, two main groups of which are available: the plunger (usually pneumatic) type (Fig. 4.25), and the screw type (Fig. 4.26). The purpose of both

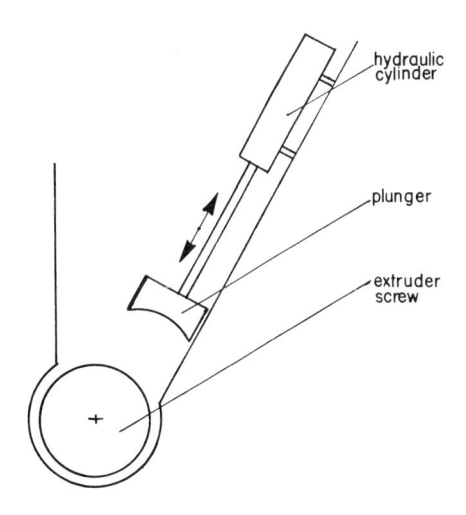

FIG. 4.25 Plunger-type stuffer.

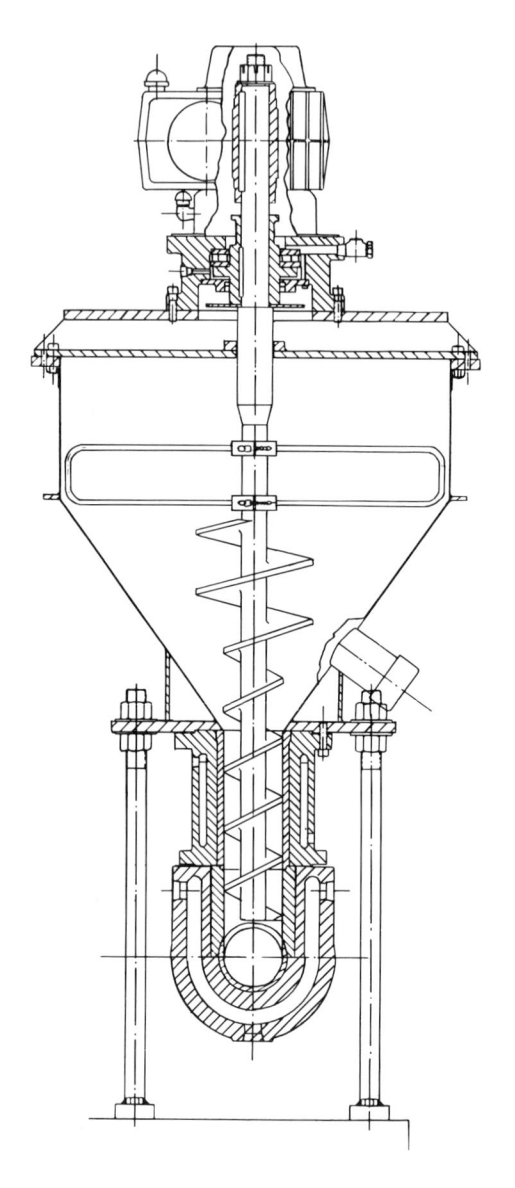

FIG. 4.26 Screw-type stuffer. (Reprinted from Ref. 12, courtesy Flag Machinery, Ltd.)

Primary Recycling

is to compress the feed material to the bulk density necessary for efficient operation of the extruder, and to force-feed the material into the extruder.

The compaction of low-density feed may also be accomplished in the feed section of the extruder. Figure 4.27 shows a schematic of a densifying extruder developed by Hartig Plastics Machinery. The diameter of the feed section of the extruder is larger than that of the rest of the screw. Without the need of a screw feeder, waste material of low bulk density is fed through an oversized feed-throat opening to the large first-stage compacting screw. This material is then compressed into the smaller-diameter plasticizing screw. The combination of different screw diameters produces the right compression ratio of low-bulk-density scrap to efficiently melt the material and feed the pelletizer or other downstream equipment [13].

Figure 4.28 shows a Condux Plastcompactor plant for processing plastic scrap material in the form of film, fibers, or foam, utilizing the heat of grinding for fusion. The waste material is shredded in the first processing unit and then conveyed via cyclone to the feed hopper. A paddle-type stirring unit in the feed hopper prevents bridging of the material, while a screw conveyor continuously conveys the material to the Plastcompactor. The centrally charged material is sintered by frictional heat between one rotating and one stationary disk, after which the material is spun off and picked up by the cold airflow from the connected central blower. Temperature control is maintained by means of a relatively low water circulation volume in the cooling chambers on the rear of the disks. (The product itself never comes into contact with the cooling medium.) The material is then pneumatically conveyed to the second cutting mill, where the sintered product is cut into a regular granulate size [14].

Figure 4.29 shows the Pallman Plast-Agglomerator, which is based on somewhat similar principles. Low-density material is ground on a knife mill and fed into a hopper. A constantly rotating horizontal stirrer loosens the feed material and provides

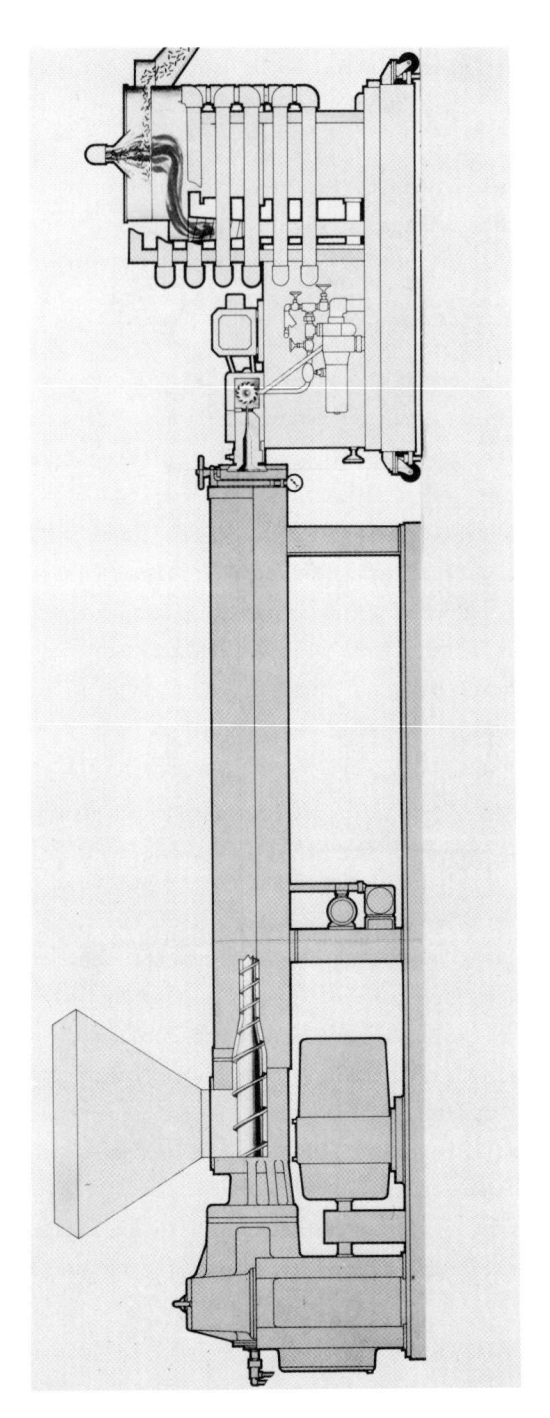

FIG. 4.27 Hartig densifying extruder. (Reprinted from Ref. 13, courtesy Hartig Plastics Machinery.)

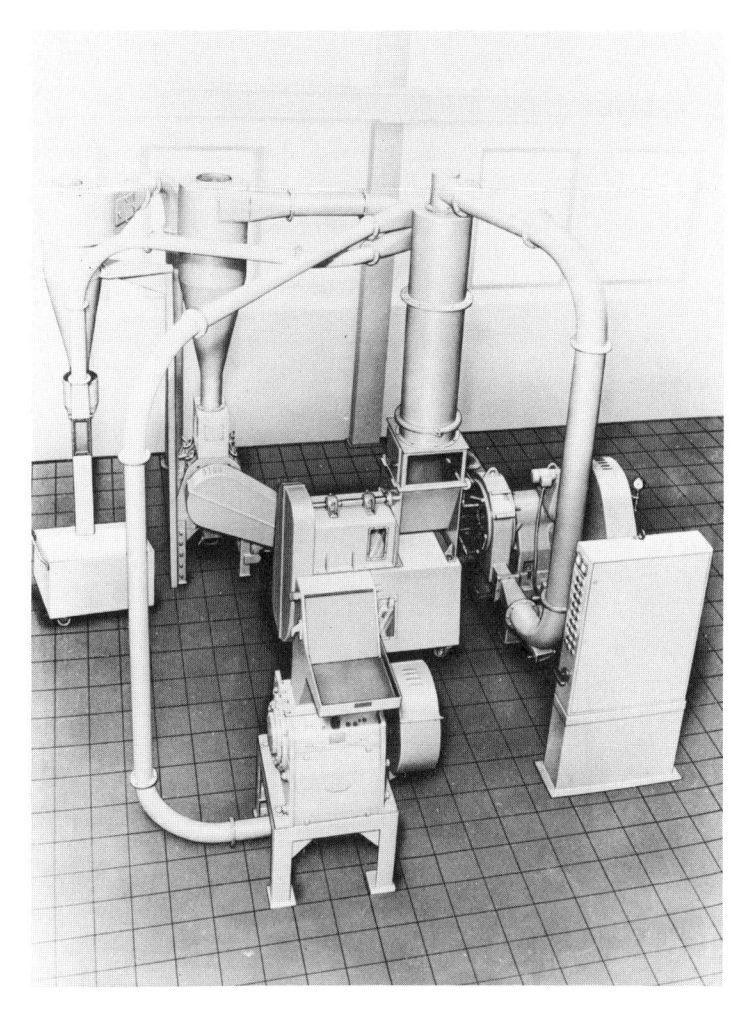

FIG. 4.28 Condux Plast-Agglomerator. (Reprinted from Ref. 15, courtesy Pallman Machinenfabrik, GmbH & Co. KG.)

for an easy flow. The feed screw transports material into the Plast-Agglomerator, where it is softened by frictional heat, and compacted and pressed through a ring with special holes. The pressed strings are cooled by suction air and cut to the desired length by a rotating knife. The product is conveyed to the

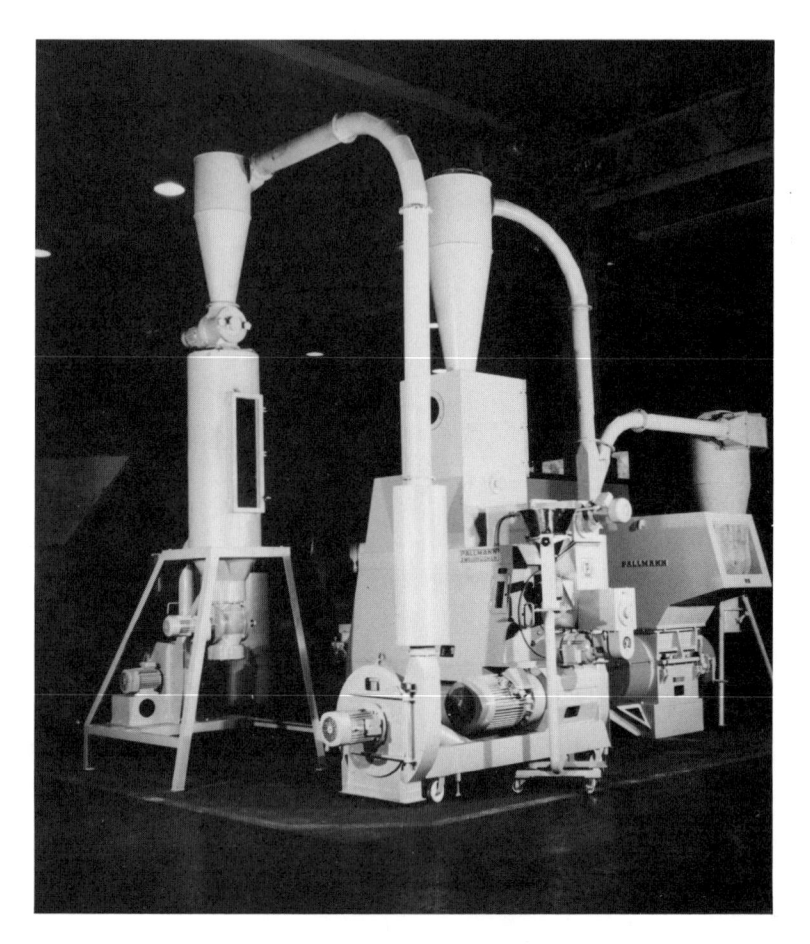

FIG. 4.29 Pallman Plast-Agglomerator. (Reprinted from Ref. 15, courtesy Pallman Machinenfabrik, GmbH & Co. KG.)

granulator where uniform granules are produced which are then conveyed to the filling station by a fan. Tangled fibers, film chips, and fines, which are not sufficiently agglomerated, are separated by gravity and recycled to the feeding hopper [15].

Figure 4.30 shows a batch film scrap densifier manufactured by Weist Industries Inc. A preweighed batch of loose film scrap is loaded into the machine and the cycle is started by the operator. The material is reduced to flakes by the rotary blade at the bottom

Primary Recycling

FIG. 4.30 Weist Industries' Vortex Automatic. (Courtesy Weist Industries, Inc.)

of the drum, and then turned into a homogeneous semiplasticized mass by the heat produced by the action. At this time water is introduced into the drum, quenching the material, which is broken into granular form by the rotor blade and automatically discharged. Total time from feed to discharge ranges from 3 to 4 min. This type of densifier is used when the relatively small amount of scrap does not justify operation of the continuous system [16].

C. In-Line Automatic Recycling Systems

Figure 4.31 is the flow diagram of a typical in-line automatic recycling system. Processing equipment (extruder, injection molding machine) produces a product and some waste (sprues, edge trim, etc.). The waste is separated from the product and conveyed to the grinder. If necessary, the ground material is densified. To assure uniform

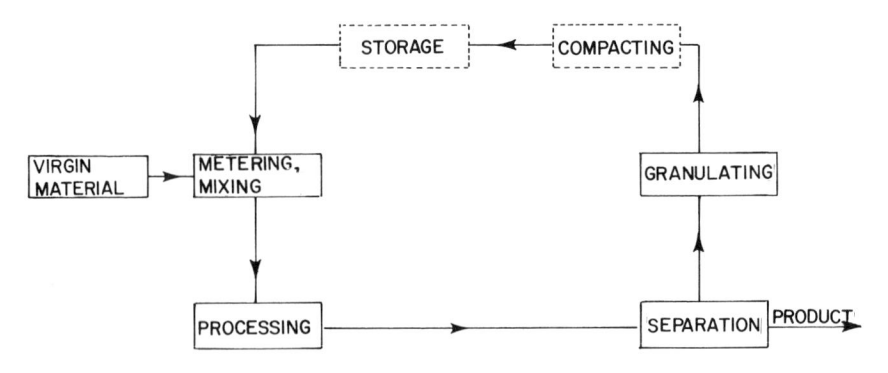

FIG. 4.31 Flow diagram of in-line recycling.

feed, storage bins may be installed. From storage, material is conveyed into the metering and mixing device where it is blended with virgin resin at a predetermined ratio. The mixture is then fed into the processing equipment.

The Automatic Scrap Recycle System (ASR) manufactured by Process Control Corporation (Fig. 4.32) is a good example of an automated in-line recycling system suitable for low-bulk-density materials. It can be used to recycle both trim and roll scrap. Edge trim is continuously pulled pneumatically from the film production line and delivered to the grinder. Roll scrap is placed on a very simple reel stand and delivered to the grinder with a roll feeder. Ground scrap is pulled from the grinder through a blower, which also serves to cool the grinder and delivered to the double compartment feed machine through the cyclone separator. The feeder is arranged to contain ground scrap in the upper compartment and virgin material in the lower (Fig. 4.33). The auger meters ground scrap at the designed rate. Virgin material flows by gravity through a clearance between the auger blades and the lower tube wall and combines with the scrap in the lower tube. The scrap is kept separated from the base material until the two are combined. The auger feeds them into the extruder. The clearance between the auger blades and the lower tube wall is great enough to allow virgin material to flow by gravity at a rate equal to the maximum extruder throughput. If the

Primary Recycling

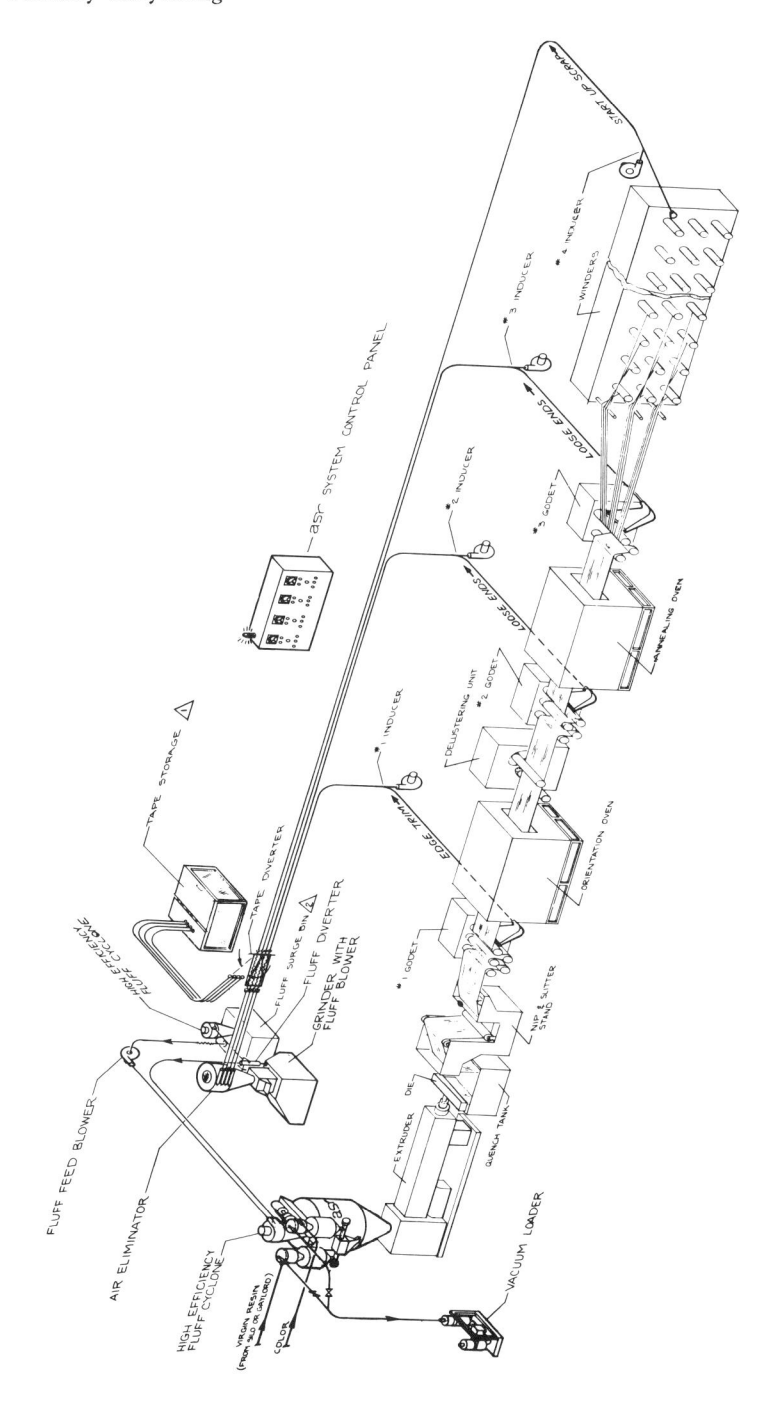

FIG. 4.32 Process Control Corporation's automatic scrap recycling system (ASR). (Reprinted from Ref. 18, courtesy Process Control Corp.)

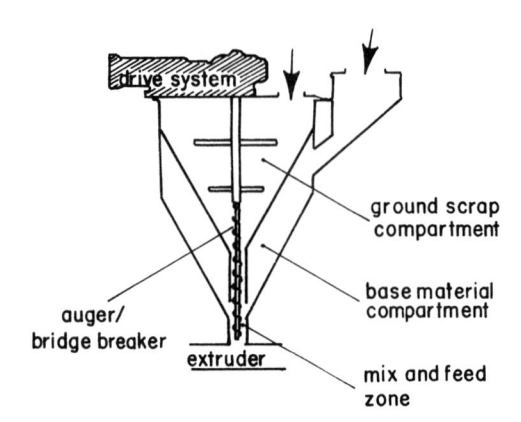

FIG. 4.33 The feeder of the Automatic Recycle System. (Reprinted from Ref. 17, courtesy Technical Association of the Pulp and Paper Industry.)

feed of regrind stops, virgin material will automatically fill the entire extruder demand. The extrusion process may thus continue regardless of variations in the availability of regrind [17, 18]. Other automatic recycling systems are available from various manufacturers, or may be assembled from available components by the processor.

REFERENCES

1. K. B. Abbas, A. B. Knutsson, and S. H. Berglund, "New Thermoplastics from Old," *Chemtech*, Aug. 1978.
2. G. Scott, "Some Chemical Problems in the Recycling of Plastics," *Res. Rec. Cons. 1:381*, 1976.
3. "Recycling and Recovery of Plastics," *Europlastics 46(3)*, 1973.
4. Y. C. Yen, "Effect of Scrap on the Mechanical Properties of Injection Moulding Parts," *35th ANTEC SPE*, 1977.
5. N. R. Schott, L. Lak, and G. Smoluk, "Recycle Calculations in Plasticating Extrusion," *32nd ANTEC SPE*, 1974.
6. A. Kelly and W. R. Tyson, *High Strength Materials*, Wiley, New York, 1965.

Primary Recycling

7. H. W. Yang, R. Farris, and J. C. Chien, "Study of the Effect of Regrinding on the Cumulative Damage to the Mechanical Properties of Fiber-Reinforced Nylon 66," *J. Appl. Polym. Sci. 23(11)*, 1979.

8. M. H. Naitove, "Buying Granulators No More a Casual Matter," *Plast. Technol. 21(5)*, 1975.

9. "Pallman Knife Mills," Brochure K421E, Pallman Machinenfabrik GmBH & Co. KG.

10. "Cryogenic Grinding Gets a Lift from New Stress on Cost Reduction," *Mod. Plast. 54(12)*, 1977.

11. N. B. Frable, "Keep Scrap Quality High with Cryogenic Grinding," *Plast. Eng. 32(5)*, 1976.

12. "Screw Feeder Series Models SF-15, 25, 35, 45, 60," Flag Machinery Ltd.

13. "Hartig and the Art of Making Pellets," Hartig Plastics Machinery.

14. "CONDUX Plastcompactor," Condux Werk.

15. "Pallman Plast-Agglomerator," Brochure K470F, Pallman Machinenfabrik GmBH & Co. KG.

16. "Model 50-500 Densilator," Foremost Machine Builders, Inc.

17. J. D. Robertson and G. L. Tedder, "Automatic Scrap Recycle System for One Step Recycling of Film Scrap," *TAPPI Paper Synthetics Conference,* Atlanta, 1975.

18. "Automatic Scrap Recycle System" Brochure G40, Process Control Corporation.

Chapter 5

SECONDARY RECYCLING

I. APPROACHES TO SECONDARY RECYCLING

Secondary recycling utilizes plastics waste unsuitable for direct reprocessing using standard plastic processing equipment. Despite the publicity received over the last few years, secondary recycling is still in its infancy. There are four main reasons for its slow development. (1) Waste plastics tend to be highly contaminated with nonplastic substances (metals, sand) posing a danger to the processing equipment; (2) various plastics present in the waste mixture used as feedstock might be mutually incompatible, resulting in a product having poor mechanical properties; (3) a feedstock with a consistent and reproducible composition is not always available; and (3) in order to be economically viable, the product must be mass produced.

Plastics wastes of various origins can be considered potential feedstock for secondary recycling processes. From the processing viewpoint these wastes can be classified four ways.

Postconsumer Plastics Waste Recovered from Municipal Refuse. This consists of a relatively consistent mixture of generic groups of plastics with a large nonplastic component. [The majority of work on the utilization of plastics from municipal solid refuse has employed simulated plastics mixtures (mixtures of virgin resins having the same composition as plastics in the refuse) or actual plastics waste handpicked from refuse. It is rarely mentioned

Secondary Recycling 157

that this approach is only of theoretical value, as most techniques used to separate plastics from refuse result in a plastic fraction the composition of which is considerably different from that of the plastics in the refuse. Most dry separation methods, for example, result in a plastic fraction containing mainly films, and thus considerably richer in polyethylene and with much better properties than the total plastic portion of the refuse.]

Postconsumer Plastics Waste Obtained from Returnable Packages. Among these are milk jars and soft drink bottles. This material consists usually of only one type of plastic and contains only small amounts of nonplastic contaminants.

Mixed Industrial Plastics Waste. This feedstock is usually obtained by collecting plastic waste from a number of industrial sources. Various plastics are present in the mixture, and the amount of nonplastic materials is small. The composition can vary with time.

Industrial Plastics Waste Consisting of a Single Type of Plastic. Usually plastic waste is too contaminated with nonplastic materials or too degraded to be used in primary recycling.

Since the products of secondary recycling are competing with other low-cost products (such as timber), secondary recycling has been developed primarily in countries where the competitive products are relatively expensive. Presently, Japan is probably the leader in secondary recycling technology, followed closely by the countries of Western Europe. Various technical approaches to secondary recycling are possible, including:

1. Reprocessing using slightly modified standard plastics processing equipment. This has the advantage of a ready availability of equipment but the disadvantages of frequent production problems and poor product properties.
2. Reprocessing using specialized processing equipment. The advantages are fast production rates and a product with

reasonable mechanical properties, while a common disadvantage is high capital cost.
3. Chemical modification of mixed plastics waste. The advantage is a product with good mechanical properties; the disadvantage is that material costs are increased without solving the processing problems.
4. Use of plastic waste in combination with virgin plastic (i.e. as a core in sandwich structure). This has the advantage that good products can be manufactured at low material cost, and the disadvantage that only certain types of relatively uncontaminated plastic waste can be used.
5. Use of plastic waste filler in other plastic or nonplastic materials. This has an advantage in that waste material is used to extend a more expensive material and a disadvantage in that the applications and types of potential products are limited.
6. Use of plastic waste as a matrix in combination with low-cost filler. The advantage is that plastic waste acts only as a binder, the mechanical properties being contributed mainly by the filler. There is a disadvantage in that the applications and types of potential products are limited.

Out of the six approaches just listed, only numbers 2, reprocessing using specialized equipment, 4, the use of plastics waste as a core in sandwich structures, and 5, the use of pulverized wastes as fillers in plastics, have been commercialized.

II. SECONDARY RECYCLING BY MECHANICAL REWORKING OF PLASTICS WASTE

A. Parameters Affecting the Properties of Blends of Noncompatible Plastics

The mechanical reworking of plastics waste utilizes the processing characteristics of waste thermoplastics and specially designed processing equipment to manufacture new products. In most cases the feedstock contains a mixture of various plastic and nonplastic materials. Approaches to the manufacture of products from mixed and contaminated waste plastics include melt homogenization, compression of the molten ground plastics without mixing, and fine grinding followed by pelletizing. Out of these three approaches, the first is thus far the most commercially successful. In order to achieve adequate mechanical properties in the resulting product, the mixture has to be thoroughly homogenized. Although the importance of dispersion uniformity and particle size is widely acknowledged, there are very few data in the literature regarding the relationship between particle size distribution and the mechanical properties of the blend. Patterson and Noah [1] studied this relationship in PE/PP and PE/PS mixtures. Both of the studied mixtures contained 5% ethylene-propylene rubber. The degree of dispersion was varied by using different mixing times and intensities. Particle size was shown to have the most pronounced influence on ultimate elongation (Figs. 5.1 and 5.2) and impact strength (Figs. 5.3 and 5.4). Decreasing particle size of the dispersed phase improves the toughness of the blend. Tensile strength generally shows similar but less pronounced dependence on the particle size of the dispersed phase. Stiffness of the blend is virtually independent of particle size [1]. There is usually a limiting particle size (1 to 5 µm) which results in maximum properties; further reduction in particle size results in no further improvement.

Paul et al. [2] studied the properties of simulated postconsumer plastics waste mixtures. Table 5.1 gives a description of the materials used in that study. The samples were prepared by melt blending in a Brabender Plasticorder, following which the hot melt was transferred to a compression mold where the test plaques were formed. The results are shown in Figs. 5.5 to 5.11. The modulus of the blends is approximately additive and can be predicted from the empirical equation:

$$E_b \approx E_1 V_1 + E_2 V_2 + E_3 V_3 + \cdots \tag{1}$$

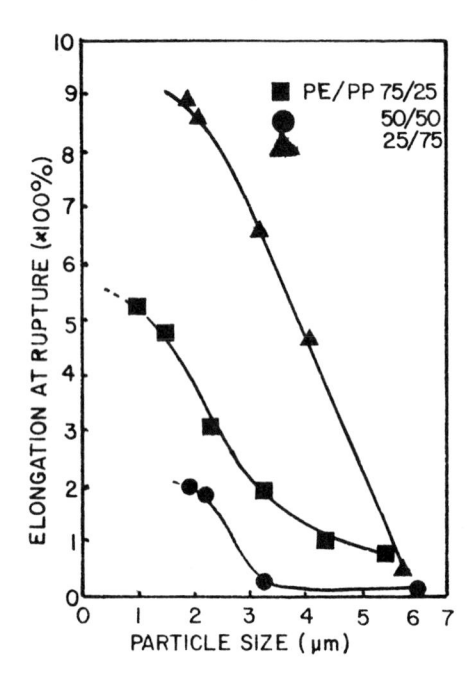

FIG. 5.1 Elongation at rupture as a function of mean particle size for PE/PP mixture. (Reprinted from Ref. 1, courtesy Society of Plastics Engineers, Inc.)

where E_b is the Young's modulus of the blend, E_1, E_2, and E_3 are the moduli of individual components, and V_1, V_2, and V_3 are the volume fractions of the components.

The other properties (tensile strength, elongation, and energy to break) are not additives. They all decrease as one moves away from any apex. A plot of any of these properties versus blend composition shows a minimum.

No blend evaluated in that work had properties better than the pure component with the highest value of that property. As expected, the variations in grade of individual components had a pronounced effect on the properties of the blend. Plasticizer in PVC results in a weaker blend with higher elongation. If one moves along any of the binary legs of the graphs toward the mid-

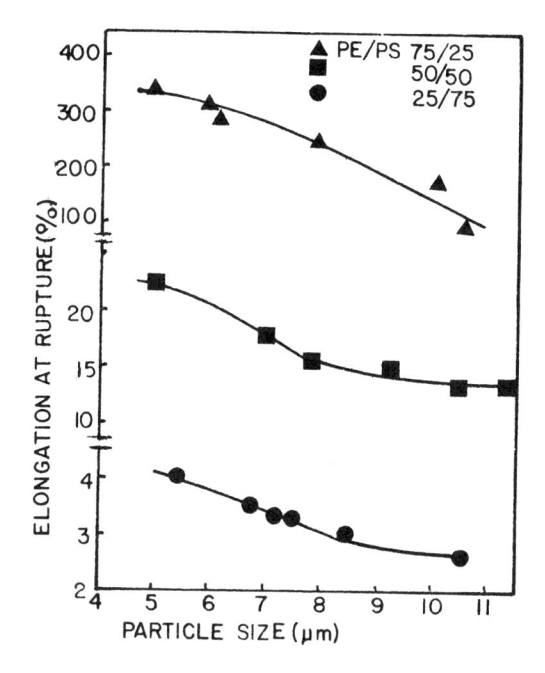

FIG. 5.2 Ultimate elongation as a function of mean particle size for PE/PS mixture (Reprinted from Ref. 1, courtesy of Society of Plastics Engineers, Inc.)

points, the properties decrease. The actual minimum does not lie exactly in the midpoint but tends to be shifted toward the component showing lower value of the given property. As the composition moves to the interior of a triangle, the properties decrease even more (for example, a blend of equal parts of PE, PVC, and PS had the lowest energy to break of all the blends).

The most important mechanical properties of simulated and actual postconsumer plastics wastes obtained by various investigators are shown in Table 5.2. Some discrepancy in the data is caused (1) by the fact that small experimental samples are not always representative (actual waste), (2) by the dependency of the properties on the grades of the component (simulated waste), and

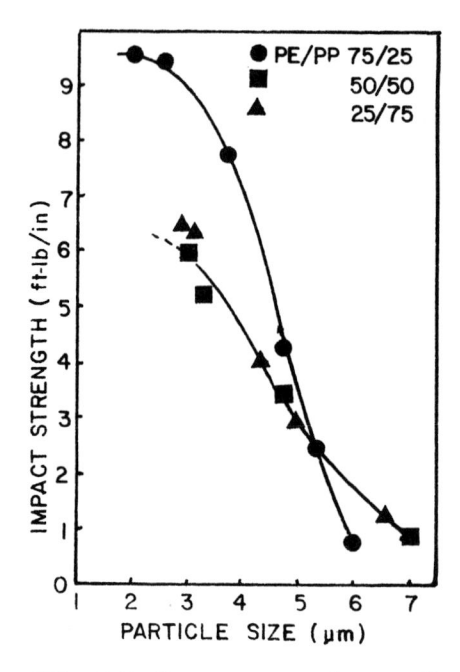

FIG. 5.3 Impact strength as a function of mean particle size for PE/PP blends. (Reprinted from Ref. 1, courtesy Society of Plastics Engineers, Inc.)

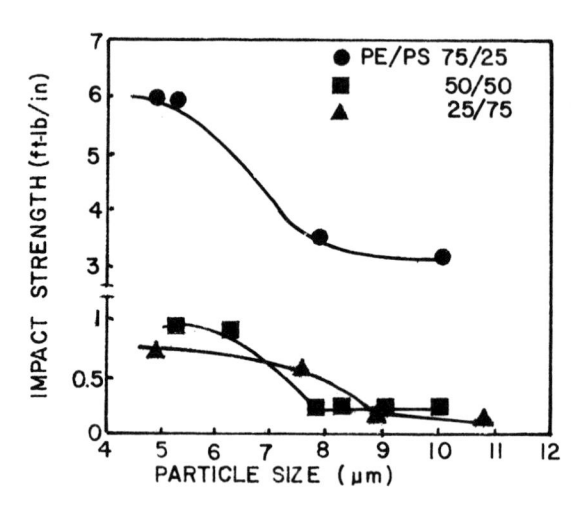

FIG. 5.4 Impact strength as a function of mean particle size for PE/PS mixtures. (Reprinted from Ref. 1, courtesy Society of Plastics Engineers, Inc.)

TABLE 5.1 Polymers Used to Prepare Simulated Postconsumer Plastics Waste

Generic type	Designation	Source	Description
Polyethylene	DYNH	Union Carbide	Den = 0.917 MI = 1.2
	DNDA	Union Carbide	Den = 0.917 MI = 20
	DNDJ	Union Carbide	Den = 0.953 MI = 4.8
	DGDA	Union Carbide	Den = 0.964 MI = 0.2
Polystyrene	Styron 685	Dow	General-purpose
	Styron 470	Dow	High-impact
Polyvinyl chloride	VYHH	Union Carbide	Medium-MW vinyl acetate copolymer
	QYSJ	Union Carbide	Low-MW homopolymer
	Geon 85542	Goodrich	Impact-modified homopolymer, blow molding compound
	K 120N	Rohm & Haas	Acrylic processing aid for PVC
	KM 611	Rohm & Haas	Acrylic impact modifier for PVC
	DOP		Dioctyl phthalate PVC plasticizer

Reprinted from Ref. 2, courtesy Society of Plastics Engineers, Inc.

(3) by the general dependency of some of the properties on the morphology of the blend. In general, mixed plastics waste is characterized by very low impact strength and elongation. Tensile strength in mixed plastics waste is also below that of competitive virgin resins.

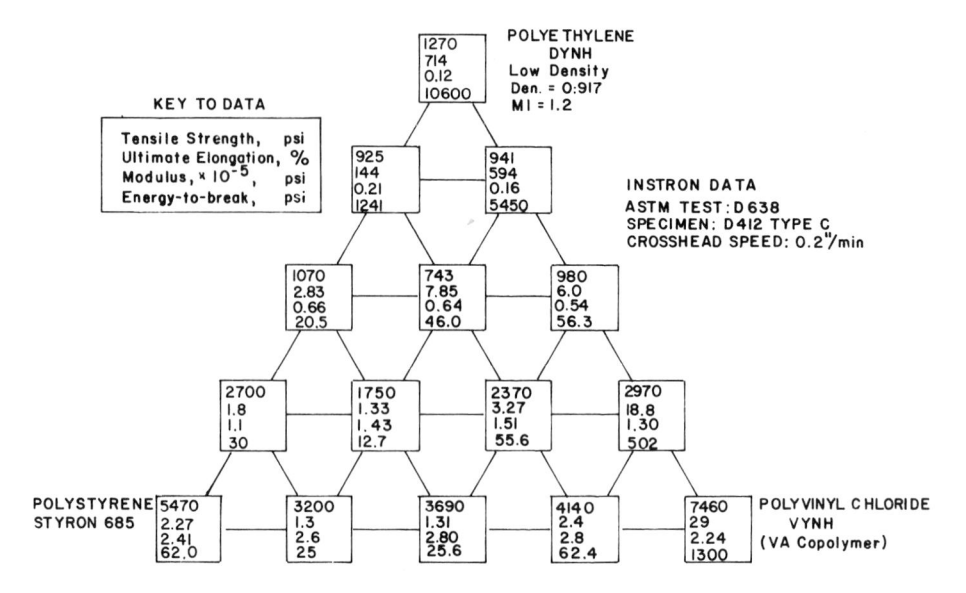

FIG. 5.5 Ternary mechanical property diagram employing low-density, low-melt-index polyethylene. (Reprinted from Ref. 2, courtesy Society of Plastics Engineers, Inc.)

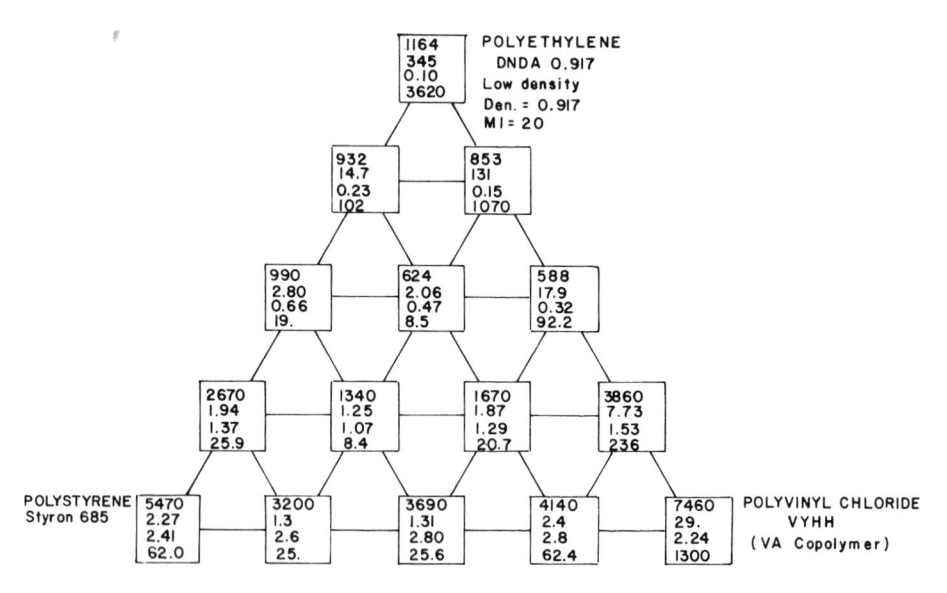

FIG. 5.6 Ternary mechanical property diagram employing high-melt-index, low-density polyethylene. (Reprinted from Ref. 2, courtesy Society of Plastics Engineers, Inc.)

Secondary Recycling

```
                    ┌──────┐
                    │ 3240 │ POLYETHYLENE
                    │ 73.5 │      DNDJ
                    │ 0.89 │  High density
                    │ 2245 │  Den. = 0.953
                    └──────┘  MI = 4.8
              ┌──────┐    ┌──────┐
              │ 2150 │    │ 2210 │
              │ 15.1 │    │ 38.9 │
              │ 0.90 │    │ 0.93 │
              │  236 │    │  568 │
              └──────┘    └──────┘
        ┌──────┐    ┌──────┐    ┌──────┐
        │ 2320 │    │ 2185 │    │ 1940 │
        │ 1.87 │    │ 1.74 │    │ 4.86 │
        │ 1.24 │    │ 1.25 │    │ 1.07 │
        │ 22.2 │    │ 19.0 │    │ 62.1 │
        └──────┘    └──────┘    └──────┘
  ┌──────┐    ┌──────┐    ┌──────┐    ┌──────┐
  │ 3930 │    │ 2710 │    │ 2720 │    │ 3510 │
  │ 3.20 │    │ 1.40 │    │ 1.77 │    │ 12.0 │
  │ 1.79 │    │ 1.93 │    │ 1.84 │    │ 1.70 │
  │ 43.2 │    │ 18.9 │    │ 26.3 │    │  345 │
  └──────┘    └──────┘    └──────┘    └──────┘
```

POLYSTYRENE 5470 3200 3690 4140 7460 POLYVINYL CHLORIDE
Styron 685   2.27 1.3   1.31 2.4   29.   VYHH
             2.41 2.6   2.80 2.8   2.24 (VA Copolymer)
             62.0 25.   25.6 62.4 1300

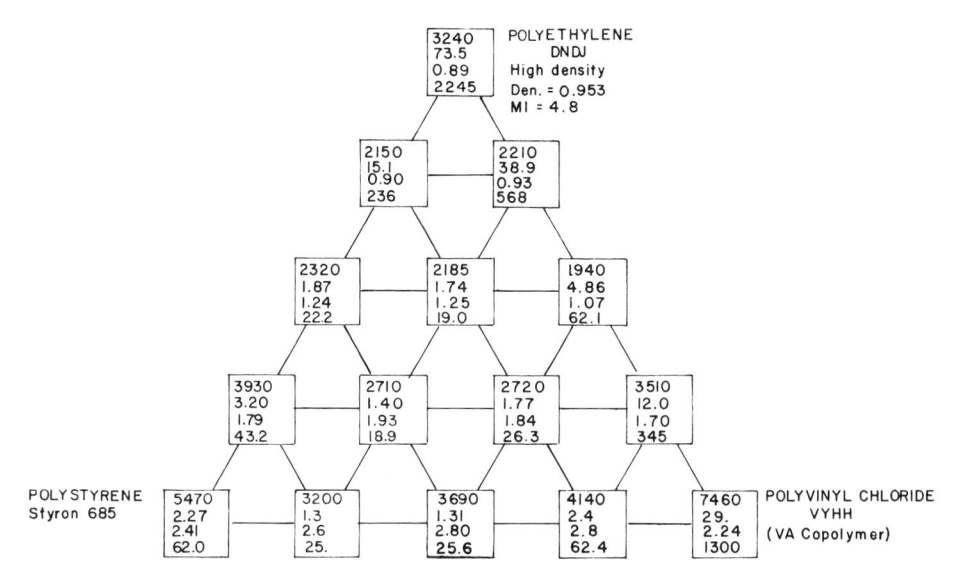

FIG. 5.7 Ternary mechanical property diagram employing high-density polyethylene. (Reprinted from Ref. 2, courtesy Society of Plastics Engineers, Inc.)

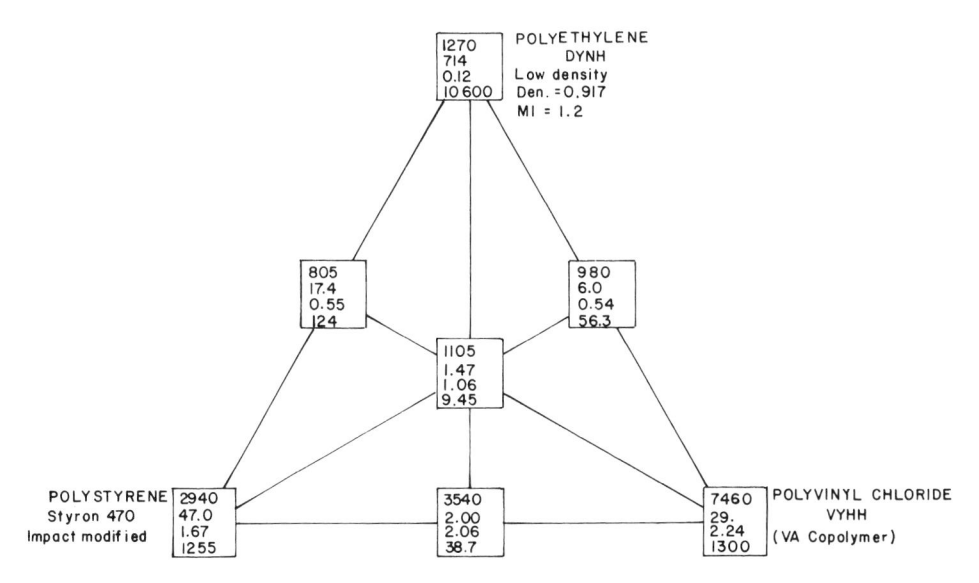

```
                    ┌───────┐
                    │  1270 │ POLYETHYLENE
                    │   714 │     DYNH
                    │  0.12 │  Low density
                    │ 10600 │  Den. = 0.917
                    └───────┘  MI = 1.2

              ┌──────┐         ┌──────┐
              │  805 │         │  980 │
              │ 17.4 │         │  6.0 │
              │ 0.55 │         │ 0.54 │
              │  124 │         │ 56.3 │
              └──────┘         └──────┘
                       ┌──────┐
                       │ 1105 │
                       │ 1.47 │
                       │ 1.06 │
                       │ 9.45 │
                       └──────┘
```

POLYSTYRENE   2940   3540   7460 POLYVINYL CHLORIDE
Styron 470   47.0   2.00   29.   VYHH
Impact modified 1.67   2.06   2.24 (VA Copolymer)
               1255   38.7   1300

FIG. 5.8 Ternary mechanical property diagram showing effect of impact-modified polystyrene. (Reprinted from Ref. 2, courtesy Society of Plastics Engineers, Inc.)

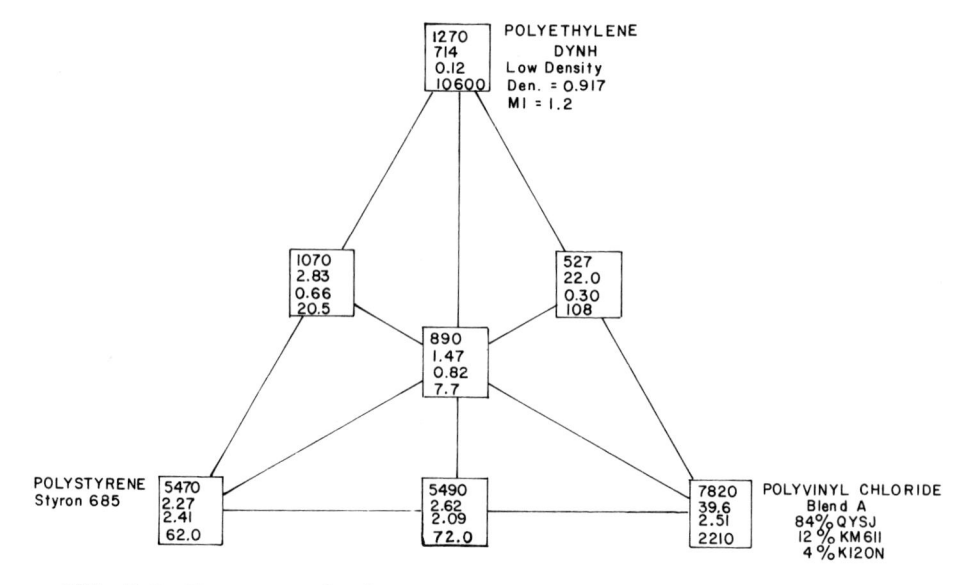

FIG. 5.9 Ternary mechanical property diagram showing effect of impact-modified PVC. (Reprinted from Ref. 2, courtesy Society of Plastics Engineers, Inc.)

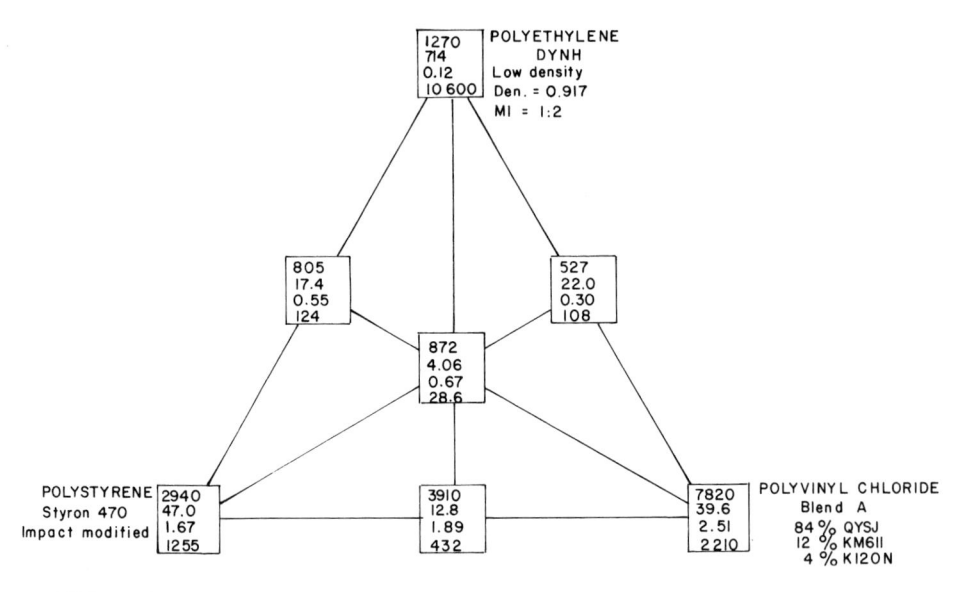

FIG. 5.10 Ternary mechanical property diagram showing the effect of both PS and PVC being impact-modified. (Reprinted from Ref. 2, courtesy Society of Plastics Engineers, Inc.)

TABLE 5.2 Mechanical Properties of Actual and Simulated Postconsumer Plastic Waste

Reference	Actual plastic waste from municipal refuse	Simulated plastic waste from municipal refuse	Tensile strength (psi)	Elongation (%)	Tensile impact strength (ft-lb/in.^{2})	Izod impact strength (ft-lb/in.)
23		X	1960	10	0	--
23	X		1450	11	0	--
31	X		1990	3		0.3

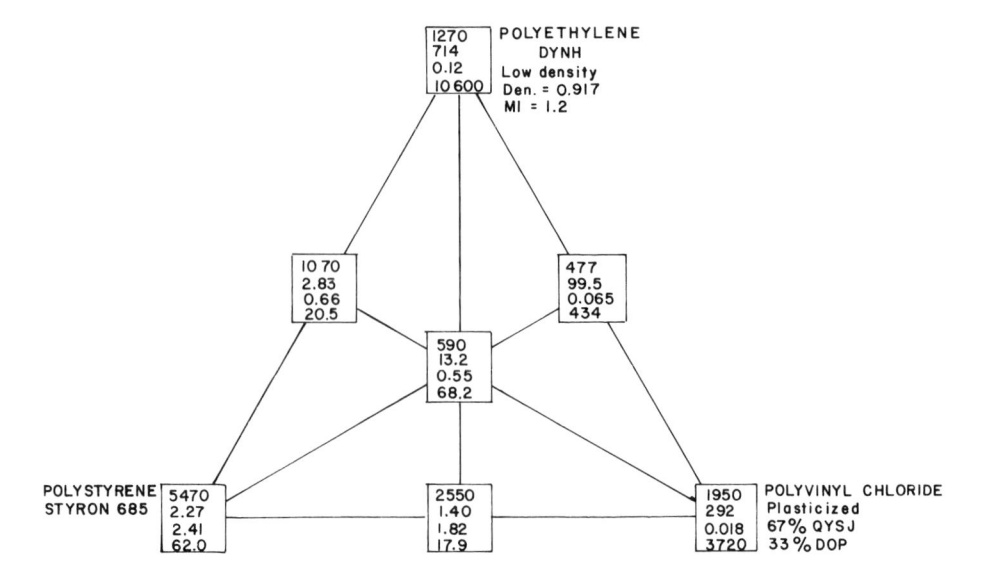

```
                    ┌──────┐
                    │ 1270 │  POLYETHYLENE
                    │  714 │     DYNH
                    │ 0.12 │  Low density
                    │10 600│  Den. = 0.917
                    └──────┘  MI = 1.2


         ┌──────┐                    ┌──────┐
         │10 70 │                    │ 477  │
         │ 2.83 │                    │ 99.5 │
         │ 0.66 │                    │0.065 │
         │ 20.5 │                    │ 434  │
         └──────┘    ┌──────┐        └──────┘
                    │ 590  │
                    │ 13.2 │
                    │ 0.55 │
                    │ 68.2 │
                    └──────┘

POLYSTYRENE ┌──────┐    ┌──────┐    ┌──────┐ POLYVINYL CHLORIDE
STYRON 685  │ 5470 │    │ 2550 │    │ 1950 │  Plasticized
            │ 2.27 │    │ 1.40 │    │  292 │  67% QYSJ
            │ 2.41 │    │ 1.82 │    │0.018 │  33% DOP
            │ 62.0 │    │ 17.9 │    │ 3720 │
            └──────┘    └──────┘    └──────┘
```

FIG. 5.11 Ternary mechanical property diagram showing the effect of plasticized PVC. (Reprinted from Ref. 2, courtesy Society of Plastics Engineers, Inc.)

B. Equipment Requirements

Smith [3] lists the following requirements for a commercially successful secondary recycling operation based on melt homogenization:

The machine has to be capable of subjecting the plastics mixture to a high shear rate at high temperature for a short time period. In order to achieve a good dispersion, high shear processing has to be used and all components of the blend must be in the molten state. During compounding, the temperature required to melt the higher-melting point-components will usually degrade components with lower melting points, which creates the need for short residence time.

The product has to be manufactured in one step. The cost of a two-step process (e.g., homogenizing and pelletizing followed by extrusion or injection molding) is usually too high to make it economically attractive.

The process should require low energy input. The recycling process should result in an energy saving as compared to the use of virgin plastics and combustion or disposal of the plastics waste.

The raw material for a given product should have a fairly constant composition. This is required to assure reproducible properties in the product.

Output must be maximized. The economics of recycling are quite different from those of manufacturing products from virgin resins. The material cost components of the total product cost is considerably lower for the recycled product, but other costs such as machine time and overhead, are considerably higher. High production rates are necessary to minimize those costs.

Proper selection of the product must be made. The mechanical properties of mixed plastics waste are usually very low, which leads to the production of parts in which thick walls are acceptable. Appearance of such parts is poor (gray color, nonuniform color, poor surface finish). Products made using mixed plastics waste cannot compete with normal products manufactured by the plastics industry. They can compete, however, with a variety of wood or concrete products when their properties are acceptable and in some cases superior to those of the original materials.

Although these requirements refer specifically to processes based on melt homogenization, they are generally applicable to other types of secondary recycling as well.

C. Commercial Equipment for Mechanical Reworking of Mixed Plastics Waste

1. The Mitsubishi Reverzer

Mitsubishi's Reverzer [4-10] is presently the most widely used process for manufacturing products from contaminated and/or mixed thermoplastics waste. It can employ as feedstock such mixed scrap as PVC, PE and nylon, and used bottles and drums. It can handle contaminated scrap such as cable stripping containing pieces of copper, bottles with paper labels, rope and string used for bailing scrap, and even sand, broken glass, and other contaminants. The following limitations on the composition of mixed feedstock apply:

The content of PS should be less than 20% to maintain good toughness.
The content of rigid PVC should be less than 50% to prevent decomposition.
Urethane can be mixed with any plastic except rigid PVC at any ratio up to 50%.

The feedstock may be further extended with up to 50% of various fillers such as sand, sludge, calcium carbonate, paper, glass fibers, and so on. The process consists basically of material preparation, extrusion, and forming. Waste plastics are pulverized in a crusher and pneumatically transported to a storage room. Here the material is dried to reduce the moisture content below 5%. From the storage room material is transported by conveyer belt to Reverzer (Fig. 5.12), which consists of a large extruder with a 90-kW motor driving a screw of about 25 cm in diameter with an L/D ratio of 3.75. On the metering end of the screw is a fluted conical section with a diameter of about 43 cm (Fig. 5.13).

As the material shears between cone and barrel, rapid melting and homogenization occur. The shear rate can be adjusted by using three friction rings to vary the gap between barrel wall and cone. Because of the rapid melting, decomposition is minimized. The melt is delivered into the accumulator, which is equipped with a degassing device to remove volatile components. A vertical screw plunger discharges the material into the mold.

During molding the material flows continuously from the extruder; the transfer system only accumulates material during the mold change and discharges it into the mold at a higher rate than the delivery from the main screw. Final products can be formed by intrusion (flow) molding, compression molding, and extrusion.

FIG. 5.12 Mitsubishi Reverzer recycling machine. (Reprinted from Ref. 6.)

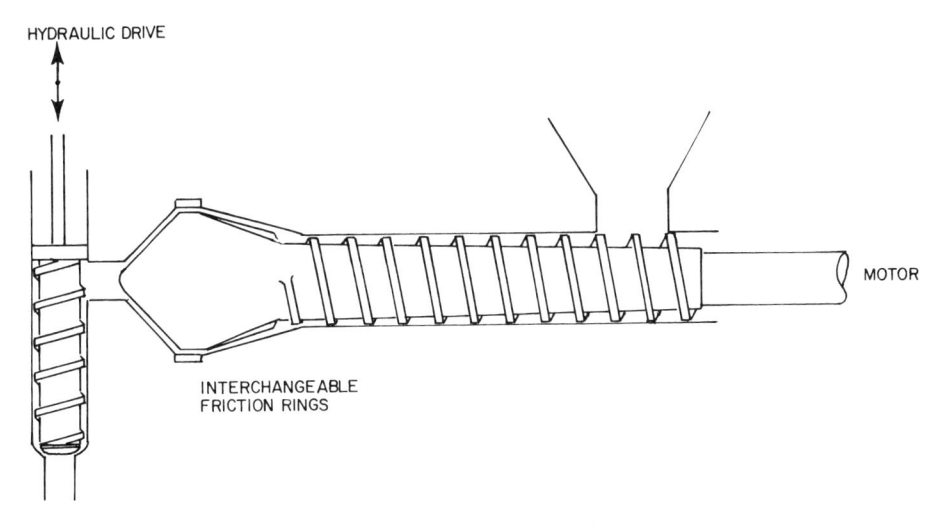

FIG. 5.13 Cross-section diagram of Reverzer. (Reprinted from Ref. 10, courtesy Rehsif S.A.)

Intrusion molding is the most widely used process. Since very low pressure is employed, inexpensive molds made of sheet metal or cast aluminum can be utilized. After filling, the molds are conveyed into a water spray tunnel for cooling and then to an unloading station where the products are removed. The empty molds are returned to the Reverzer for refilling. Usually approximately 20 molds are used. Very large objects weighing up to 45 kg have been produced using this technique.

If the product has a large projected area, or a good surface finish is required, compression molding can be used. In this process the filled molds are passed into a hydraulic press where the melt is cooled under pressure. The Reverzer can also be used to produce continuous sections by extrusion. Systems are available for the extrusion of solid and tubular profiles at rates of up to 450 kg/hr.

The output of the Reverzer system depends on the flow properties of the feedstock, the particle size after grinding, and the size of the product. The average output of the process is between 4 and 6 metric tons per day (two shifts) or between 7 and 11 metric tons per day (three shifts).

Products most suitable for the Reverzer process are large, heavy section items which can replace wood or concrete components. The cost of such moldings is less than that of comparable components manufactured from wood or concrete, but because of their superior resistance to weather and fungus attack they are sold at prices above those of wood products. Following are examples of products made by this process (Figs. 5.14 and 5.15):

Vertical posts:	2000 × 110 × 110 mm
Horizontal rails:	6000 × 100 × 45 mm
Stakes for agriculture:	1800 × 40 × 40 mm
Drainage gutters:	1500 × 300 × 300 mm
Cable reels:	630- to 1200-mm diameter
Paving blocks:	500 × 500 × 100 mm
Drain pipes:	150- × 1000-mm diameter
Cargo skids:	920 × 580 × 215 mm

FIG. 5.14 Cable reel manufactured by Reverzer process from waste plastics. (Reprinted from Ref. 6.)

FIG. 5.15 Ranch fending made by Reverzer process. (Reprinted from Ref. 6.)

2. *The Klobbie*

Figure 5.16 shows a photograph of the Klobbie, an extruderlike plastic waste-recycling machine developed by Rehsif and based on the invention by E. J. G. Klobbie of Lankhorst Touwfabrieken in the Netherlands. Because the machine handles various plastics wastes, sometimes having low bulk densities, it is equipped with a hopper containing a crammer-feeder which mixes material and forces it into the feed section of a screw. The material is homogenized and plasticized by a high speed (approximately 350 rpm) adiabatic extruder. The extruder does not require external heating as sufficient heat is generated by the shearing action of the screw. The high screw speed results in good mixing and short residence time. The extruder pumps the molten plastic into the molds. Ten

FIG. 5.16 The Klobbie recycler. (Reprinted from Ref. 11, courtesy Rehsif S.A.)

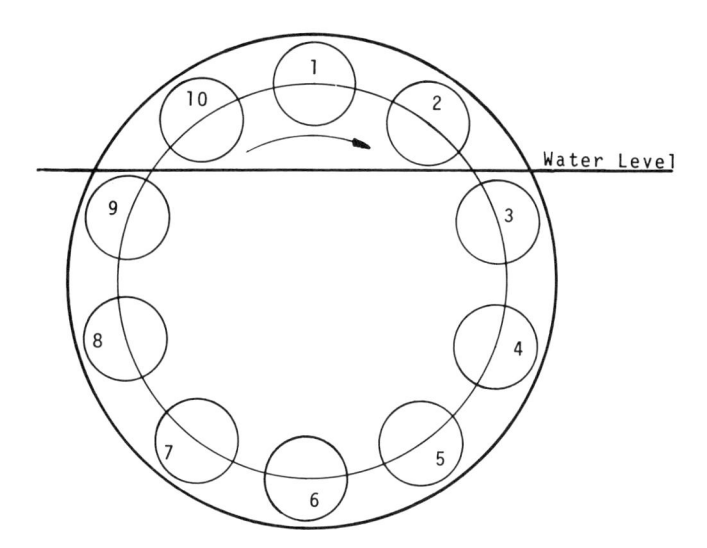

FIG. 5.17 Schematic of the mold turret: (1) filling, mold closed pneumatically; (2) mold closed pneumatically; (3) - (9) cooling, mold closed by springs; (10) ejection. (Reprinted from Ref. 11, courtesy Rehsif S.A.)

molds are used, mounted on a turret with a horizontal shaft. In order to minimize tooling costs, external water cooling is used. The molding sequence is schematically shown in Fig. 5.17. In position (1) the mold is held against the extruder-adapter by a force of 1-1/2 to 2 metric tons exerted by a pneumatic cylinder. After the mold is filled, the material emerging from the air vent at the end of mold strikes the microswitch, which stops the extruder screw, releases the mold locking pressure, rotates the mold turret by 36°, actuates the mold locking piston, starts the rotation of the extruder screw, and, immediately prior to the filling station, pneumatically ejects the molded part from the mold (10). An automated molding sequence allows the Klobbie to handle ten different molds at once. All the molds have to be of the same size, though different parts may be molded.

The Klobbie is being used to manufacture products such as piles for harbor works, canal bank and coast erosion protection, electric cattle fences, regular fences, and marker posts for highways and car parks. Such products can be handled with normal woodworking tools, but unlike wood, they are rotproof, fungus resistant, insectproof, and resistant to seawater [11,12].

3. The FN Machine

The FN machine resulted from Professor Patfoort's experiments on Weissenberg effect machines carried out at CRIF (Centre des Recherches Scientifiques de l'Industrie des Fabrications Métalliques) in Liège. The development work was completed by Fabrique Nationale Herstal near Liège. The machine, shown schematically in Fig. 5.18, uses a short screw (L/D = 5) with the end cut off at right angles to the axis. The flat front rotates across the stationary bottom surface of the barrel. The plastic material is compacted at the barrel wall and then melted by being subjected to shear in the space ahead of the screw front surface. The two main advantages of this method of plasticization are (1) a very short time period during which the material is in the molten state; and (2) an automatic control of melt viscosity, since particles with a higher viscosity than average are subjected to

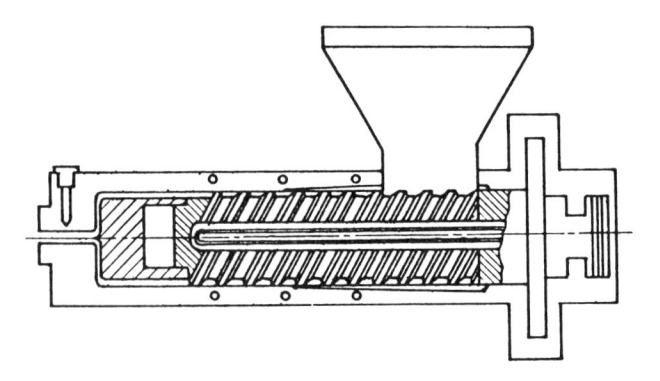

FIG. 5.18 Schematic of the FN machine. (Reprinted from Ref. 135, courtesy Kunststoffe.)

Secondary Recycling 177

greater shear, resulting in higher temperatures which in turn lower the melt viscosity. The FN process produces pellets that can be converted into final products on standard plastics-processing equipment [12,13].

4. The Flita System

The Flita system, designed by Flita GmbH of West Germany, utilizes a roller mixer eccentrically mounted in a cylindrical mixing chamber. The material is melted by externally applied heat and homogenized by being sheared against the wall of the mixing chamber. The products are manufactured by compression molding in separate presses [12].

5. The Remaker

Kleindienst in Germany is manufacturing small intrusion molding machines for recycling relatively clean industrial plastic waste [12,14,15]. The machine is shown in Fig. 5.19. The machine contains a specially designed plasticizing screw in which a spiked drum revolves in a box studded internally with spikes. The machine accepts a wide variety of plastics waste. Commercial products have been manufactured with rigid or flexible PE, PP, ABS, and PS. Mixed plastic bottles with various types of decorations still attached to them have been successfully recycled by the unit. Because of relatively long residence time in the heated cylinder, the problems might be encountered in handling mixtures containing PVC.

Since low pressures are used, only inexpensive aluminum tooling is required. Relatively large parts weighing between 1 and 2 kg can be produced on the Remaker. The best results are achieved with heavy, thick-walled objects with large cross sections and parts that do not require very close dimensional tolerance. Table 5.3 lists examples of the products manufactured on the Remaker, some of which are shown in Fig. 5.20.

FIG. 5.19 Remaker. (Reprinted with permission from ASTM STP 533. Copyright ASTM, 1916 Race Street, Philadelphia, PA, 19103.)

FIG. 5.20 Typical parts molded on Remaker. (Reprinted with permission from ASTM STP 533. Copyright ASTM, 1916 Race Street, Philadelphia, PA, 19103.)

TABLE 5.3 Typical Products Manufactured on Remaker

Item	Cavities	Shot weight (oz)	Cycle (sec)	Feedstock
Shoe soles	2	19	50	PVC
Luggage handles	4	7	45	PE, PS, ABS
Suction pump	1	6	60	PVC
Seals	4	5	50	PVC
Toy animals	6	4	45	PS, PVC
Bicycle saddles	1	12	90	PVC
Bicycle pedals	4	8	60	PE, PVC
Lawn mower wheels	1	9	45	PE, PVC
Door stops	4	7	50	PE, PVC, PS
Dog toys	4	10	60	PVC, PE, PS
Cable housing	2	24	90	PP, PE
Auto parts	Various	2 to 8	30 to 60	PP, PVC

Reprinted with permission from ASTM STP 533. Copyright ASTM, 1916 Race Street, Philadelphia, PA, 19103.

6. The Regal Converter

The design of equipment for recycling mixed and contaminated waste plastics can be considerably simplified if the need for mixing and homogenizing the melt is eliminated. This approach has been used in the "Converter" developed by Regal Packaging.

The process accepts various mixed and contaminated thermoplastic wastes with up to 50% paper, wood, dust, and so on. The waste is fed into the granulator where relatively uniform chips are produced. A pneumatic system conveys the granules to the Converter (Fig. 5.21). The granules are passed through the oven on a steel belt, where the particles melt and fuse together. The material is then passed through the roller arrangement where it undergoes compaction. Continuous board 1.32 m wide by 2 to 20 mm thick can be produced at a speed of up to 1 m/min. The hot and still soft sheet is transferred to the hydraulic press where the final product is formed between the cold platens. The properties of the board depend on the materials, and can vary from rigid to flexible. It is also possible to form a sandwich of virgin resin/waste plastics/virgin resin structure. The products are waterproof and can be used for fencing (Fig. 5.22), boxes, low-cost pallets (Fig. 5.23), shrink wrap trays, and so on [16-19].

FIG. 5.21 Converter. (Reprinted from Ref. 16, courtesy Plastic Recycling, Ltd.)

Secondary Recycling 181

FIG. 5.22 Enclosure made on Converter. (Reprinted from Ref. 16, courtesy Plastic Recycling, Ltd.)

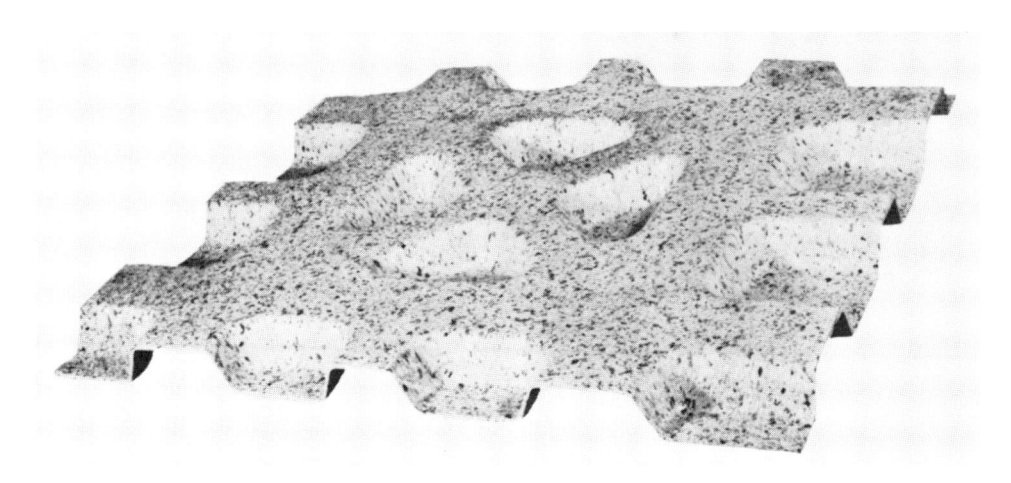

FIG. 5.23 Pallet made on Converter. Reprinted from Ref. 16, courtesy Plastic Recycling, Ltd.)

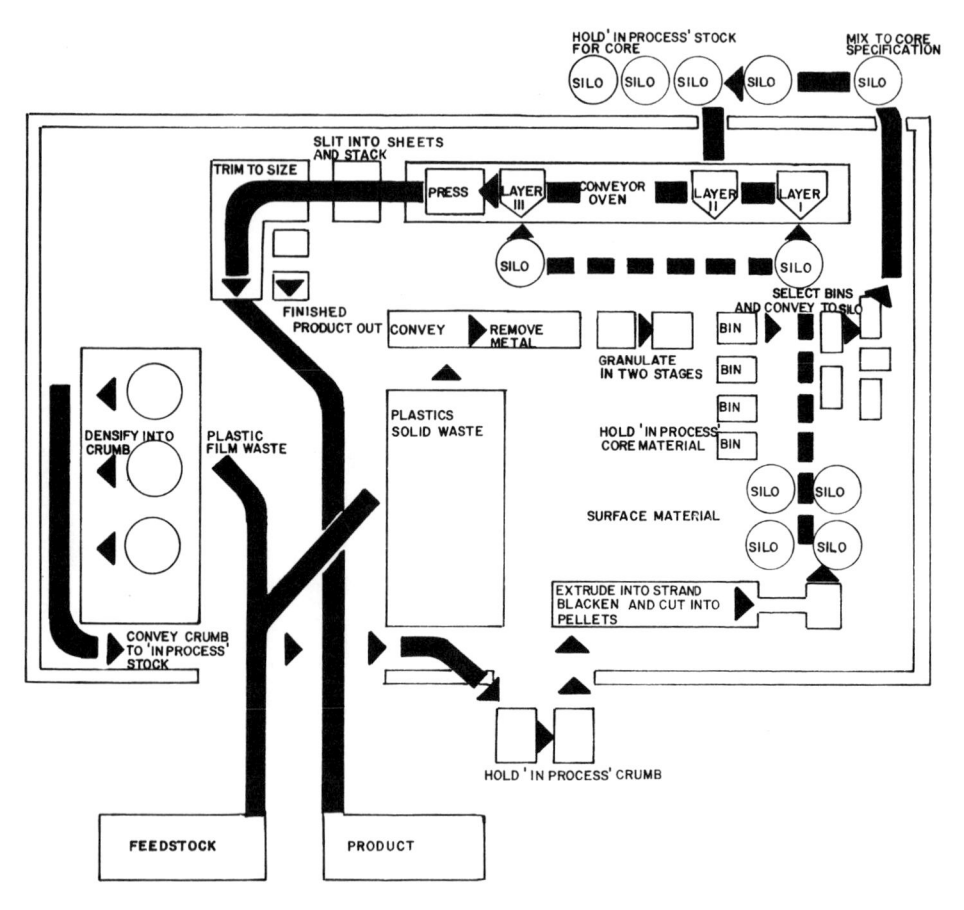

FIG. 5.24 Flowsheet of Tufbord manufacturing line. (Reprinted from Ref. 20, courtesy Reclamat International Limited.)

7. *Reclamat Tufbord*

The process developed by Reclamat International Limited is somewhat similar to that developed by Regal [20]. Figure 5.24 is a flowsheet of the process. Two feed materials are used, waste film and solid waste plastic. Waste plastic film is pulverized and densified to form small crumbs which are compounded with carbon black in an extruder equipped with a pelletizing die. Pellets are deposited on the heated steel belt where they undergo fusion. The second feed material is also granulated and deposited on the heated steel belt. The two materials combine to form a laminate having skins made of recycled film material and core made of recycled solid plastic. The layers are joined together on an 1800-metric-ton hydraulic press linked to the belt in such a way that the belt stops while the pressing takes place. A continuous sheet 1.2 m wide by 9 to 12 mm thick can be produced; it is easily sewn, nailed, and screwed with conventional tools or welded together with a hot-air gun and PE welding rods. The product has good impact resistance and does not absorb water. The black surface gives the appearance of uniformity and protects the product from the effects of UV radiation. The applications include cladding, partitioning, fencing, and paneling.

8. Kabor's K Board

Kabor Limited developed a sheet extruded from polyethylene-coated paper [8,21]. This feedstock is supplemented by extra polyethylene. K Board was mainly directed toward the farm market, its applications including paneling for farm buildings (Fig. 5.25), bottom paneling for feeding troughs, and floor coverings. The company found sources of suitable PE scrap of adequate quality difficult to acquire, and went out of business in 1974.

9. *Japan Steel Works Limited, Nikko Waste Plastics Reclamation Line*

A demonstration plastics recycling plant capable of processing about 10 tons of plastics waste per week was set up in Japan by Japan Steel Works Limited [22]. Figure 5.26 shows a schematic of that process. Plastics wastes separated by the householder are delivered to the plant by truck. Bags containing waste are automatically unloaded and fed by conveyer into a crusher which cuts the material into approximately 5-cm pieces. In order to protect the crusher from large pieces of metal, each bag with waste plastics is automatically weighed; those weighing over 0.9 kg

FIG. 5.25 Farm building in Lincolnshire, England, with K Board for its side walls. (Reprinted from Ref. 21.)

are rejected. The shredded waste is given a preliminary wash, after which it is conveyed to an air classifier. A second crusher grinds the plastics waste into very small particles, which are then blown through a magnetic selector that further eliminates ferrous metals. The material is washed in water and detergent solution, and then dried. It is fed into a turbo mill where it is ground even finer, and then conveyed to an extruder where pellets are produced. The pellets consist of about 40% LDPE, 10% HDPE, 10% PP, 15% PS, 15% PVC, and 8% thermosets. The remaining 2% is made up of nonplastic materials. The mechanical properties of the mixture are very poor: its tensile strength is only 2.5 kg/cm^2. The pellets are used to make flower pots by injection molding.

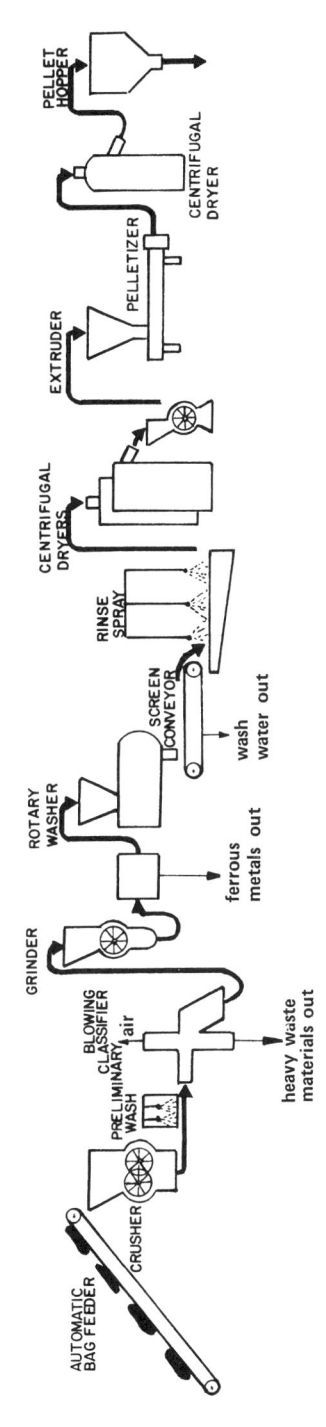

FIG. 5.26 Schematic of Japan Steel Works' Nikko Waste Plastics Reclamation line. (Reprinted from Ref. 22, courtesy McGraw-Hill.)

III. CHEMICAL MODIFICATION OF MIXED PLASTICS WASTE

Mechanical mixtures of incompatible plastics, such as those recovered from municipal refuse, have inherently poor mechanical properties. Lack of adhesion between various phases results in stresses being carried mainly by the continuous phase. The dispersed phase not only does not contribute to the strength of the blend, but in some cases causes stress concentration, which further reduces the mechanical properties. A considerable amount of work has been reported on the upgrading of mechanical properties of mixed plastics waste through an improvement in interfacial bonding. The two basic approaches suggested are addition of compatibilizers and cross-linking of the blend.

A. Upgrading the Properties of Mixed Plastics Waste by Use of Compatibilizers

Plastics present in municipal refuse consist mainly of PVC, PE, and PS. All of these materials are mutually incompatible. The ideal compatibilizer for any two of these plastics would be a long-chain polymer one end of which was compatible with one phase and the other end of which was compatible with the other. Schram and Blanchard proposed chlorinated PE (CPE) as a possible PE/PVC compatibilizer [23]. Work on such systems has been carried out by researchers at the University of Texas [24,25]. The chlorinated PE used in these studies was produced by the chlorination of HDPE particles suspended in water. In the reaction chlorine placement occurs only in the amorphous phase. The product has a blocklike structure, some segments of the molecular chain of which are similar to PE and others of which are similar to PVC. Polymers of this structure are expected to have surfactantlike characteristics which allow them to locate preferentially at the domain boundaries in a blend of incompatible polymers (PE and PVC); thus, the bonding is increased between the domains.

Table 5.4 describes polymers and Table 5.5 describes CPE modifiers used in Ref. 24. Figures 5.27, 5.28, and 5.29 show the effects of the addition of CPE into ternary blends of PE, PVC, and PS on the elongation at break, yield tensile strength, modulus, and energy to break, respectively. The original blend showed only 1.4% elongation. Addition of 30% of CPE 36 or CPE 42 results in an increase in tensile elongation of almost 100%. CPE 48 is somewhat less effective in that respect. Both tensile strength and Young's modulus decrease with the addition of CPE. In spite of these decreases, energy to break (a measure of a toughness of the material) shows an increase similar to the increase in elongation. Results obtained by adding CPE to the actual scrap plastic mix also show considerable improvement in the properties (Table 5.6). The most pronounced change was observed in impact strength. Addition of 27.5% of CPE increased tensile impact strength from virtually 0 to 2.83 ft-lb/in^2. It was

TABLE 5.4 Polymers Used at the University of Texas Study

Generic type	Designation	Source	Description
Polyethylene	DYNH	Union Carbide	Den = 0.917 g/cm^3 MI = 1.2
	DGDA	Union Carbide	Den = 0.964 g/cm^3 MI = 0.2
Polystyrene	Styron 685	Dow	General-purpose
Polyvinyl chloride	VYHH	Union Carbide	Medium-MW vinyl acetate copolymer
	Geon 85542	Goodrich	Impact-modified homopolymer, blow molding compound

Reprinted from Ref. 24, courtesy Society of Plastics Engineers, Inc.

TABLE 5.5 CPE Modifiers

Property	CPE 36	CPE 42	CPE 48
Chlorine (%)	36	42	48
Relative crystallinity (%)	8	8	0
100% Modulus (psi)	150	250	225
Tensile strength (psi)	1400	2100	2100
% Elongation	800	550	450

Note: All data and materials were supplied by the Dow Chemical Co. CPE 36 and CPE 48 correspond to Dow sales designations of CPE 3623 and CPE 4814, respectively. CPE 42 is an experimental polymer with the designation XP 2243.31.

Reprinted from Ref. 24, courtesy Society of Plastics Engineers, Inc.

shown in the subsequent study [25] that CPE also changes blend morphology. Figure 5.30 shows the effect of adding 20% CPE 36 to the blends of PE and PVC. The blend containing CPE has a considerably finer structure. Unexpectedly, it was found that CPE also changes morphology of binary PE/PS mixtures.

TABLE 5.6 Physical Properties of Actual Plastics Waste with Increasing Levels of CPE 42

CPE/scrap plastic mix (%)	Elongation (%)	Tensile (psi)	Tensile impact ft-lb/in.^{2})
0/100	11	1450	0
15/85	11.7	1715	0.45
17.5/82.5	12.7	1690	0.54
20/80	15.7	1715	0.76
22.5/77.5	17.7	1712	1.5
25/75	20	1600	1.6
27.5/72.5	22	1600	2.83

Reprinted from Ref. 23, courtesy Society of Plastics Engineers, Inc.

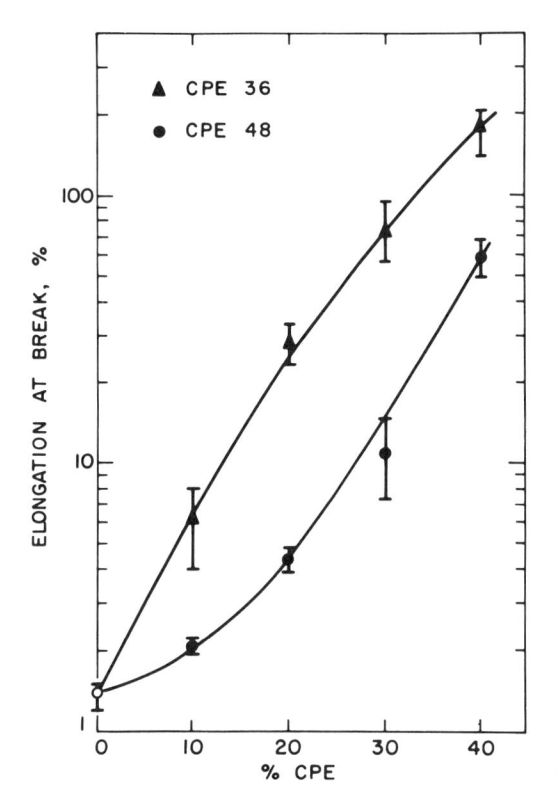

FIG. 5.27 Elongation at break for ternary blends composed of equal parts by weight of DYNH/VYNH/Styron 685; %CPE is based on total weight. Data for CPE 42 are identical with those for CPE 36. (Reprinted from Ref. 24, courtesy Society of Plastics Engineers, Inc.)

Ram et al. studied the effects of various polymeric additives on PE/PS blends [26]. It was shown that additives such as EPDM or EVA can increase tensile strength and impact resistance of these blends. Locke and Paul [27,28] studied the effect of PE/PS graft copolymers on PE/PS blends. Graft copolymers were made by impregnating PE with styrene and then grafting styrene onto PE by irradiation. Grafts obtained with a radiation dose of 0.5 mrad appear to be optimum blend modifiers. Improvements in excess of

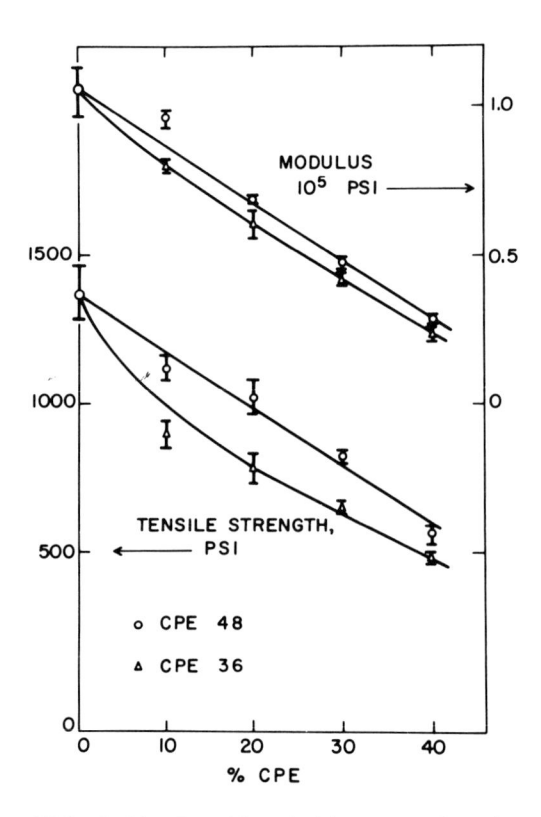

FIG. 5.28 Tensile yield strength and modulus of blends having composition as shown in Fig. 5.27. Data for CPE 48 are identical with those for CPE 42. (Reprinted from Ref. 24, courtesy Society of Plastics Engineers, Inc.)

50% for strength and 100% for elongation were obtained using 33% modifier. Similar studies [29,30] using different PS/PE graft copolymers resulted in some increase of the yield strength and elongation at yield, but also in a decrease in elongation at break.

The action of polymeric compatibilizers is often compared to that of surfactants. Both of these substances have molecular chains one end of which is soluble in the continuous phase and the other of which is in the dispersed phase. There is, however, a substantial difference. Surfactant molecules have great mobility in the

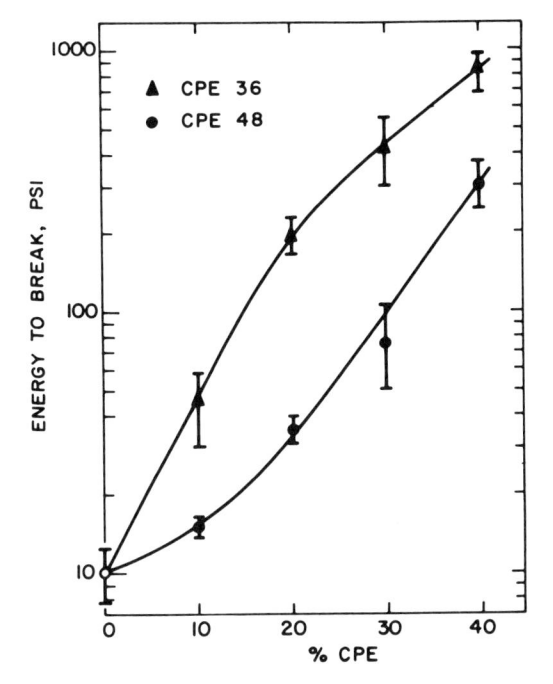

FIG. 5.29 Energy to break for ternary blends having composition as shown in Fig. 5.27. Data for CPE 42 are identical with those for CPE 36. (Reprinted from Ref. 24, courtesy Society of Plastics Engineers, Inc.)

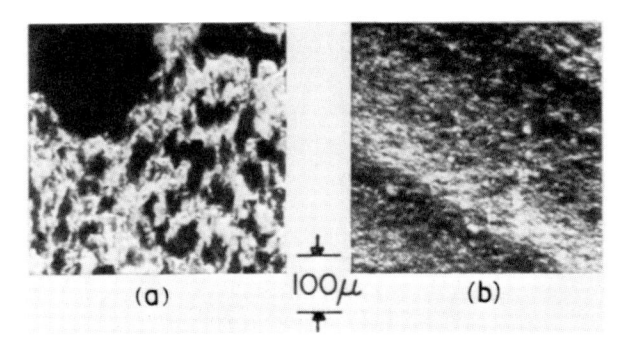

FIG. 5.30 Photomicrographs obtained using crossed polarizers. (a) 50:50 mixture of PE and PVC, (b) Same as (a) + 20% CPE 36. (Reprinted from Ref. 25, courtesy Society of Plastics Engineers, Inc.)

fluid which allows them to migrate freely to the interface. Thus only very small amounts of surfactants are needed. Very-long-molecular-chain compatibilizers in the polymer melt do not have that mobility. In order to achieve an adequate concentration at the interface, from 10 to 30% compatibilizer must be used. Although property improvements are achieved, the cost advantage of using low-cost waste materials is much diminished.

B. Chemical Cross-Linking

Chemical cross-linking using high-temperature peroxides is an effective way of improving the mechanical properties of mixed post-consumer plastics waste recovered from municipal refuse [34-36]. Polyethylene, which constitutes a considerable portion of that waste, can be cross-linked by compounding with peroxide such as dicumyl peroxide or di-t-butyl peroxide, and then heating the mixture. Cross-linking of polyethylene extends the upper temperature at which this plastic can be used; cross-linked PE is a solid at temperatures where non-cross-linked PE melts and flows. Cross-linking occurs in three steps:

1. Decomposition of the peroxide and formation of peroxy radicals:
$$ROOR \longrightarrow 2RO\cdot$$

2. Abstraction of hydrogen from PE chain:
$$RO\cdot + -CH_2CH_2- \longrightarrow ROH + CH_2\overset{\cdot}{C}H$$

3. Cross-linking by coupling of the polymer radicals:
$$2-CH_2\overset{\cdot}{C}H- \longrightarrow \begin{array}{c} -CH_2CH- \\ | \\ -CH_2CH- \end{array}$$

It is highly probable that if the mixture of plastics is compounded with the appropriate peroxide, not only will the PE phase undergo cross-linking, but other phases will also, and, what is more important, some cross-linking will occur at the interfaces. Such cross-linking would improve the mechanical properties of the blend.

Secondary Recycling

In the work described in Refs. 31-33 the actual plastics waste recovered from municipal refuse was used. Decomposition of the peroxide and the subsequent cross-linking are temperature-dependent phenomena. Figure 5.31 shows a typical torque time curve obtained using the Brabender Plasticorder. The material tested was mixed plastics waste containing 1 pph of DiCup 40C (40% dicumyl peroxide suspended on calcium carbonate, Hercules, Inc.). When cold material is charged into the mixer the torque rapidly increases. Heating and melting of the material is accompanied by a decrease in viscosity followed by an increase due to the cross-linking. Figure 5.32 shows the dependence of curing time on temperature. Too high molding temperatures will cause the formation of voids. Figure 5.33 shows a decrease in the density of cross-linked mixed plastics waste as a function of molding temperature. Due to the decomposition of certain components there is a considerable decrease in density above 320°F. If a higher temperature is used, the generation of voids can be minimized by cooling the material under pressure; the gases produced will remain dissolved in the plastic.

Figures 5.34 to 5.37 show the effects of chemical cross-linking on Young's modulus, tensile strength, tensile elongation, and Izod impact strength, respectively. Young's modulus shows a drastic decrease due to cross-linking: the cross-linked material becomes more flexible. Tensile strength decreases and tensile elongation increases dramatically due to the cross-linking. For most applications tensile elongation of 20 to 30% should be adequate. The 30% elongation achievable with approximately 3.5 pph of DiCup 40C represents a 1000% increase from the elongation of non-cross-linked plastics waste. Lower stiffness and increased elongation result in improved impact strength, which increases almost linearly with the amount of peroxide added. At a high degree of cross-linking, high impact strength values of approximately 3 ft-lb/in. are achieved. Figure 5.38 is a typical output from an instrumented impact tester comparing the impact behavior of cross-linked and non-cross-linked mixed plastic waste. It is interesting to note

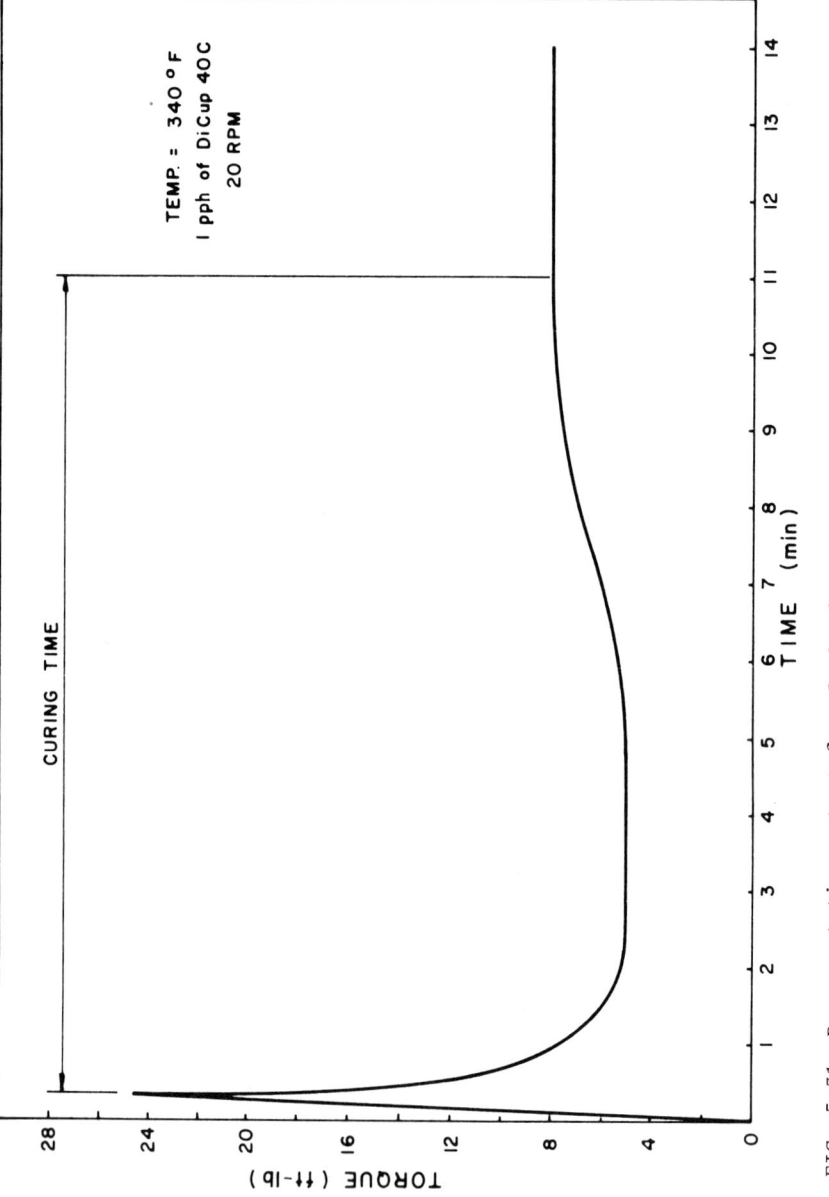

FIG. 5.31 Representative output from Brabender Plasticorder equipped with mixing head showing increase of viscosity with cross-linking of mixed waste. (Reprinted from Ref. 31, courtesy Ontario Ministry of the Environment.)

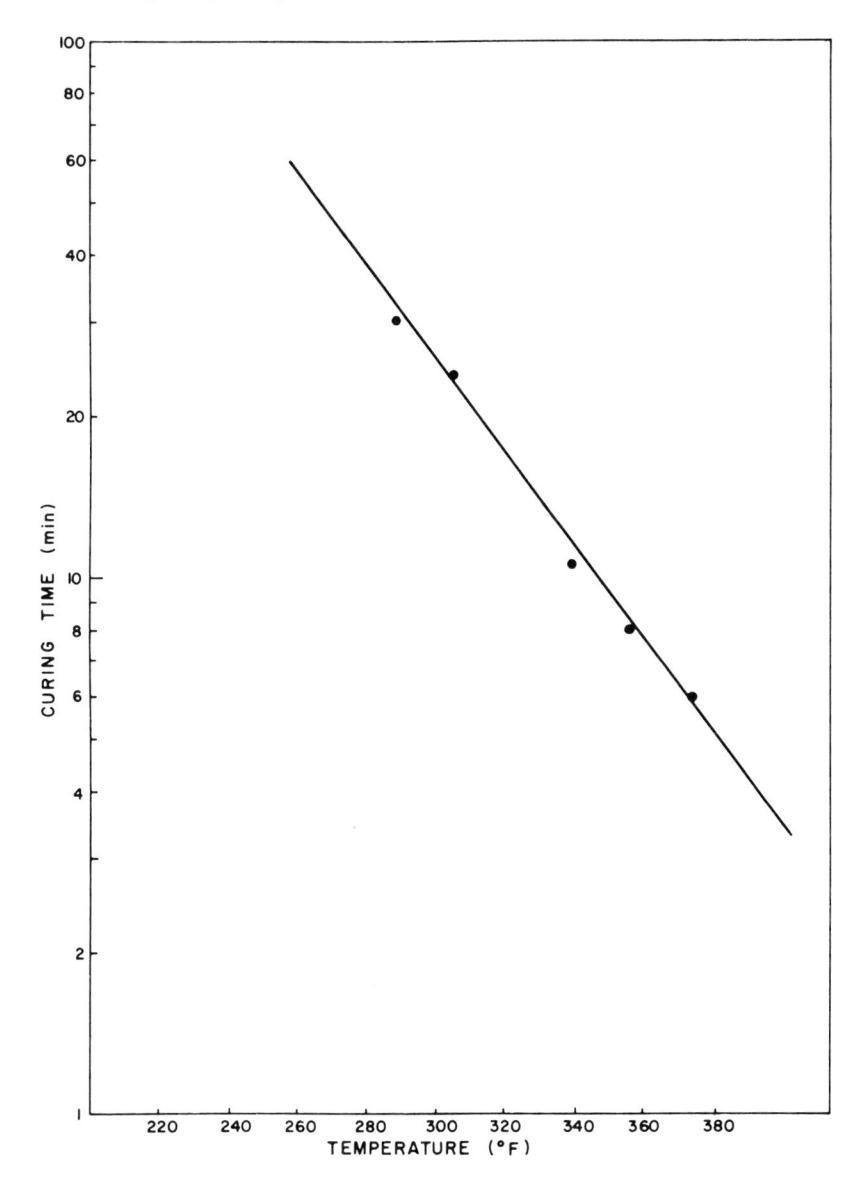

FIG. 5.32 Time necessary to cross-link mixed waste as a function of temperature. (Reprinted from Ref. 31, courtesy Ontario Ministry of the Environment.

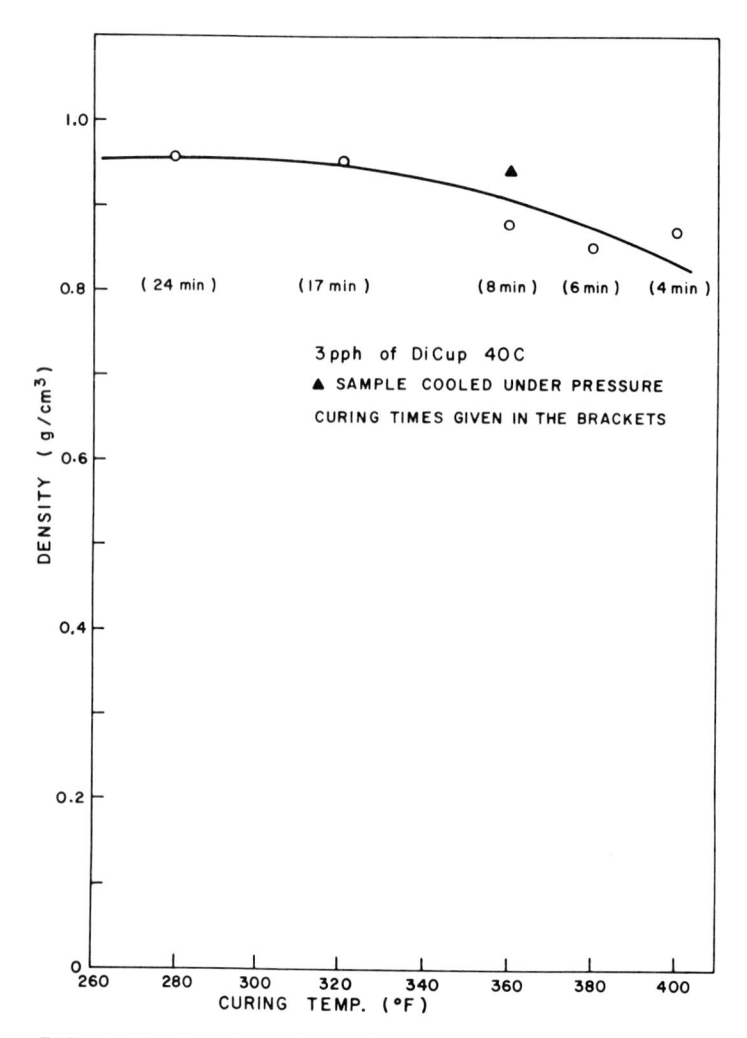

FIG. 5.33 Density of mixed plastic waste samples as a function of molding temperature. Above 320°F some components of the mixture decompose, creating voids in the samples. (Reprinted from Ref. 31, courtesy Ontario Ministry of the Environment.)

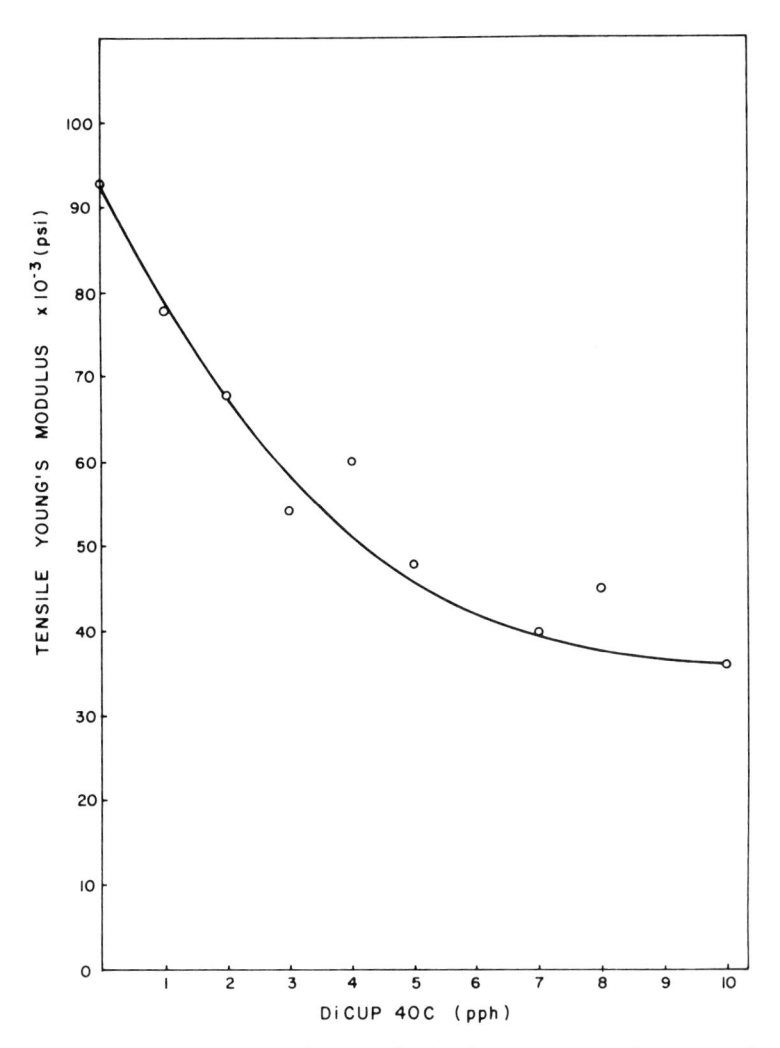

FIG. 5.34 Tangent modulus of mixed waste plastics as a function of the degree of cross-linking. Addition of cross-linking agent makes the material more flexible. (Reprinted from Ref. 31, courtesy Ontario Ministry of the Environment.)

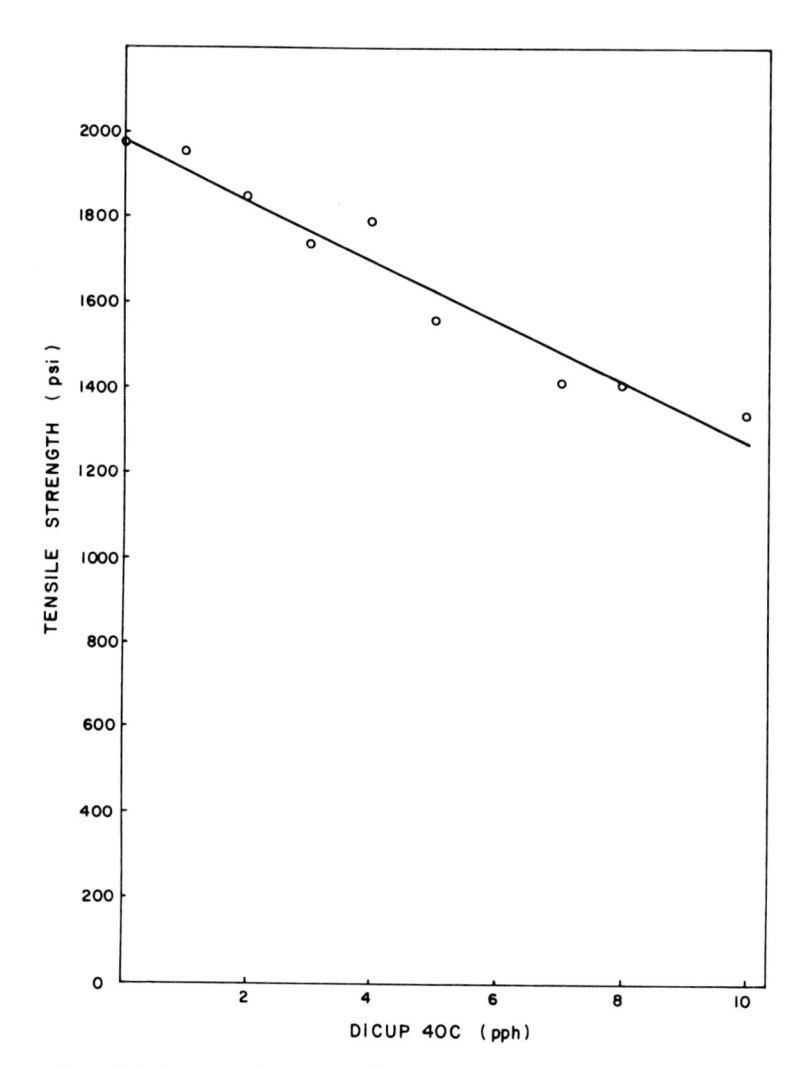

FIG. 5.35 Tensile strength of mixed waste plastics as a function of the degree of cross-linking. Cross-linking causes decrease of tensile strength. (Reprinted from Ref. 31, courtesy Ontario Ministry of the Environment.)

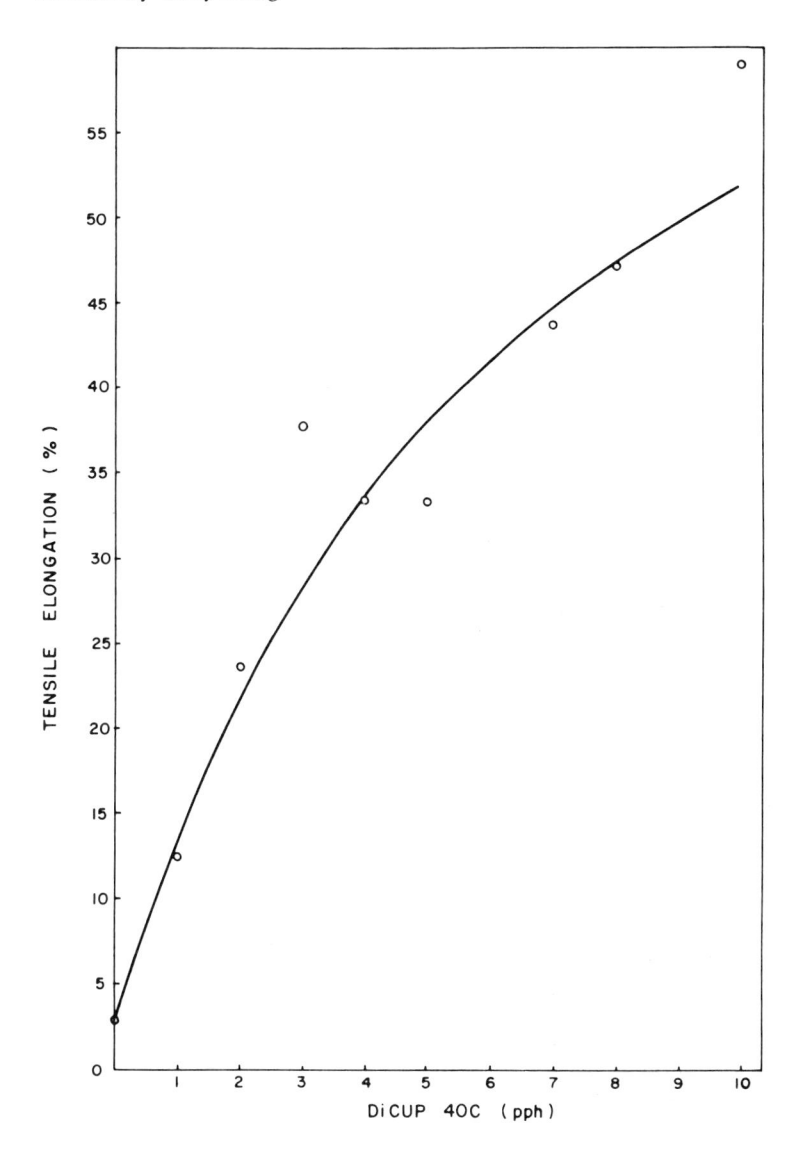

FIG. 5.36 Effect of cross-linking on tensile elongation. Eight pph of DiCup 40C increases elongation of mixed plastic waste about 15 times. (Reprinted from Ref. 31, courtesy Ontario Ministry of the Environment.)

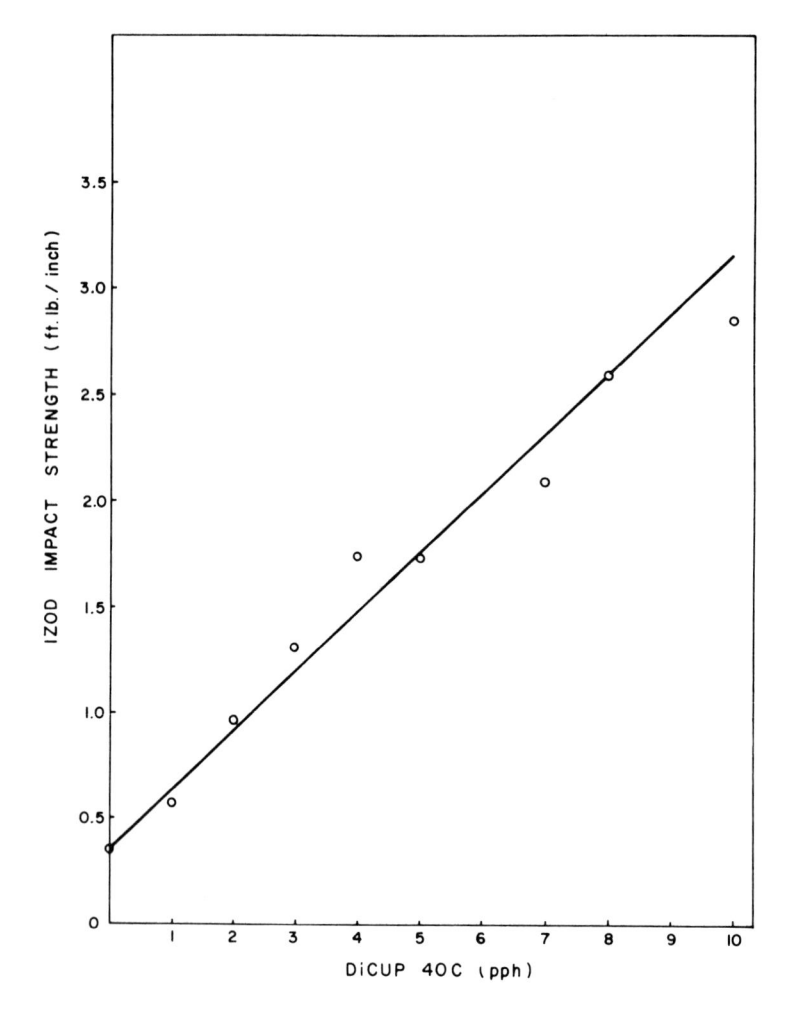

FIG. 5.37 Izod impact strength of mixed plastic waste as a function of the degree of cross-linking. High impact strength of approximately 3 ft-lb/in. can be achieved by cross-linking. (Reprinted from Ref. 31, courtesy Ontario Ministry of the Environment.)

FIG. 5.38 Representative output from instrumented impact tester comparing impact behavior of mixed plastic waste with and without cross-linking. (Reprinted from Ref. 31, courtesy Ontario Ministry of the Environment.)

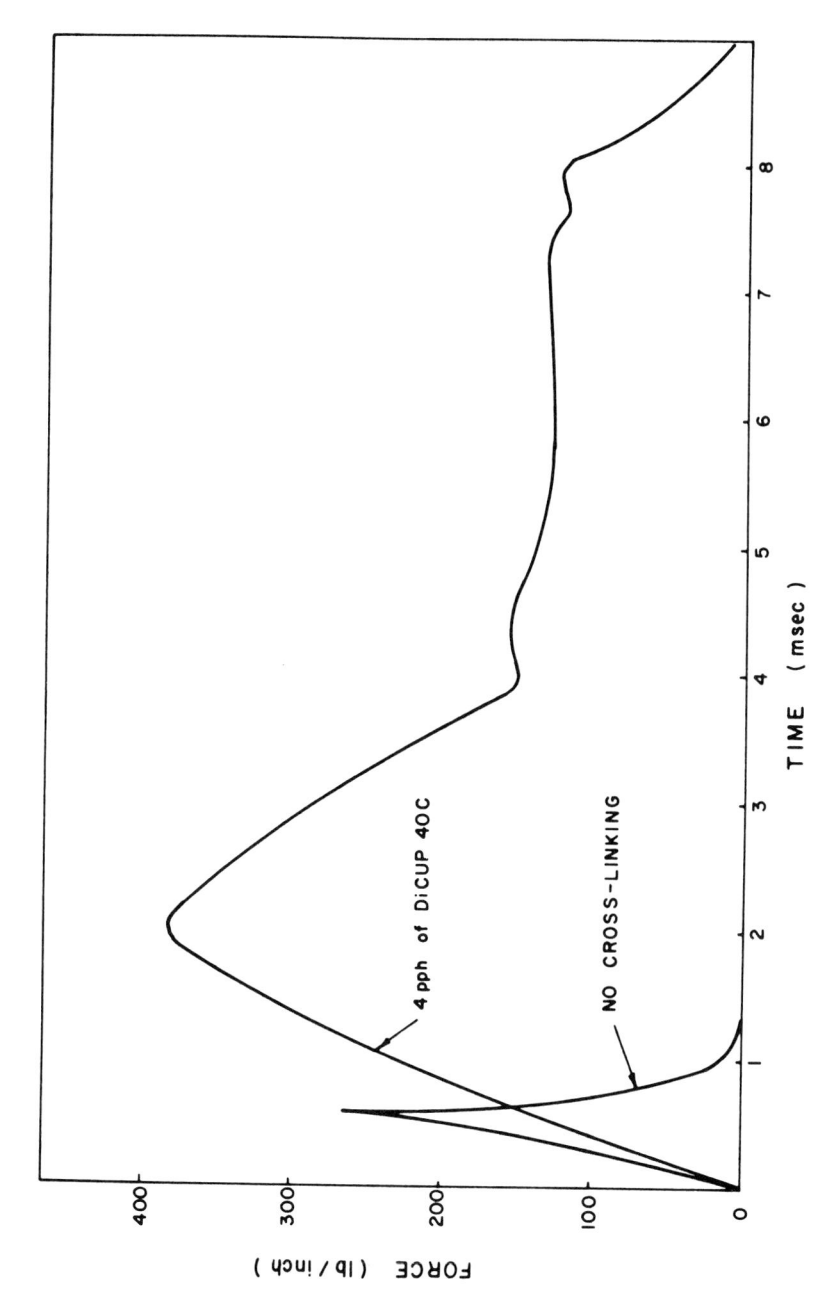

that under impact conditions the cross-linked material shows higher strength than the non-cross-linked one. Cross-linking transforms a brittle mixture of incompatible plastics into flexible, highly impact-resistant material. The presence of a considerable amount of paper reduces the toughness of the cross-linked plastics waste. The presence of 20 pph of paper reduces tensile elongation from approximately 23% to approximately 5% and Izod impact strength from approximately 1 ft-lb/in. to approximately 0.4 ft-lb/in. Young's modulus increases from 60,000 to 110,000 psi and tensile strength increases from approximately 1800 to approximately 2100 psi. The loss of toughness can be compensated for by an increased amount of peroxide [31].

References 31 and 32 also describe the preparation of foams from cross-linked mixed plastics waste. The molding was accomplished in the following steps:

Collecting plastics waste
Washing
Shredding (6-mm screen)
Melting and mixing in the extruder
Grinding
Compounding with the peroxide on the roll mill
Addition of blowing agent
Removal from the roll
Compression molding at 350°F for 18 min under approximately
 1400 psi pressure
Opening the mold and expansion of the foam
Cooling the foam

Commercial blowing agent Celogen AZ (Uniroyal Chemical) was used. Figure 5.39 is a cross section of a typical sample. The presence of various phases and of nonplastic materials resulted in nonuniform cell sizes. The density of the foam is a function of the concentration of the blowing agent (Fig. 5.40). An initial rapid decrease in density at high density ranges is followed by a

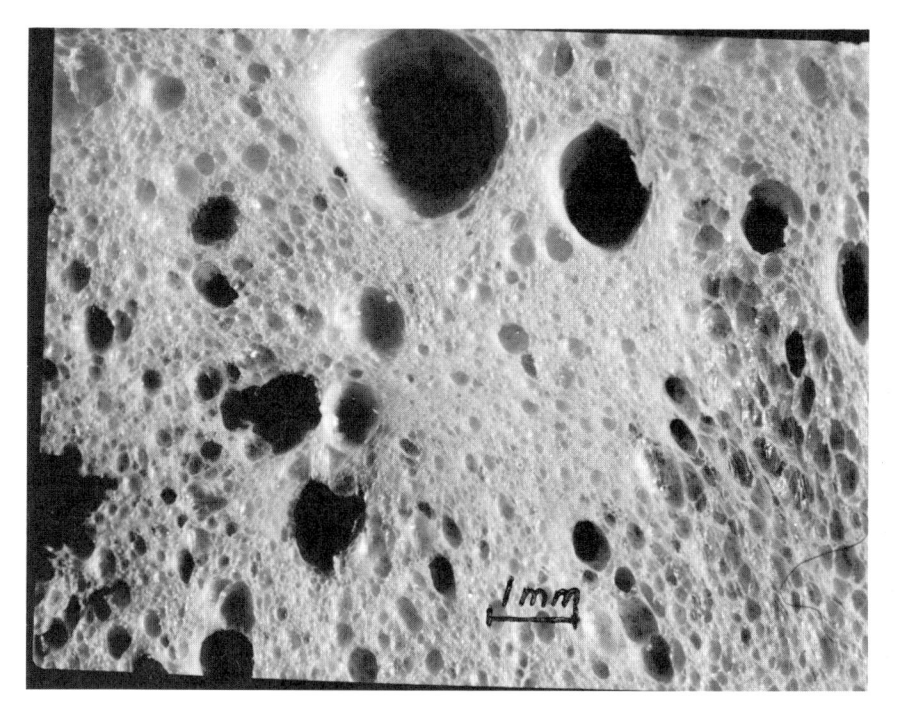

FIG. 5.39 Cross section of a typical sample of expanded cross-linked waste plastics.

small decrease in density with an increase in blowing agent at lower density ranges. The degree of cross-linking also influences the density of the foam (Fig. 5.41) and an increase in the amount of cross-linking agent reduces the foaming efficiency of the blowing agent. No successful foams have been obtained at more than 5 pph of DiCup 40C. With higher degrees of cross-linking, the material will not expand due to the internal pressure, but will fracture. Compression properties of mixed plastics waste foam are similar to those of cross-linked PE foam.

Cross-linking is an efficient method of upgrading the properties of mixed plastics waste recovered from municipal refuse. Since material processed in that fashion becomes a thermoset, it cannot

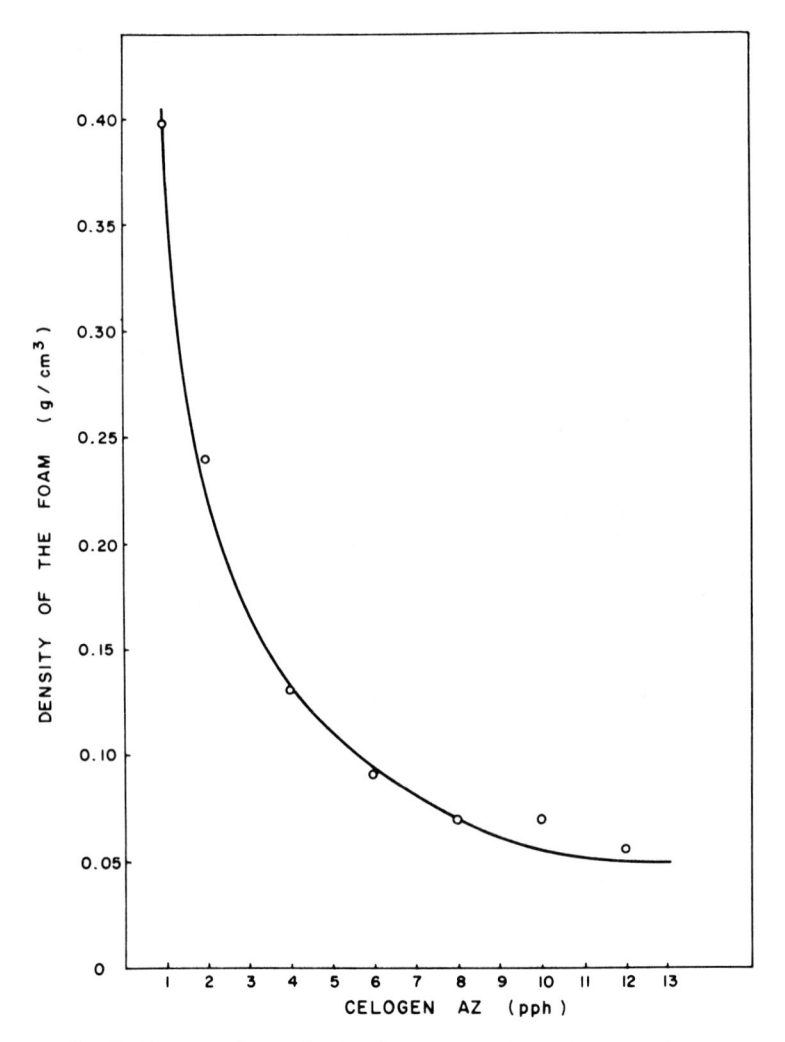

FIG. 5.40 Density of the foam as a function of the amount of blowing agent (Celogen AZ). (Reprinted from Ref. 31, courtesy Ontario Ministry of the Environment.)

be reprocessed again. Tough sheeting competing with plywood in some applications can be produced using a calendering line adapted for handling waste plastics. Extrusion followed by high-temperature curing could also be used. Since the practical levels of toughness

Secondary Recycling

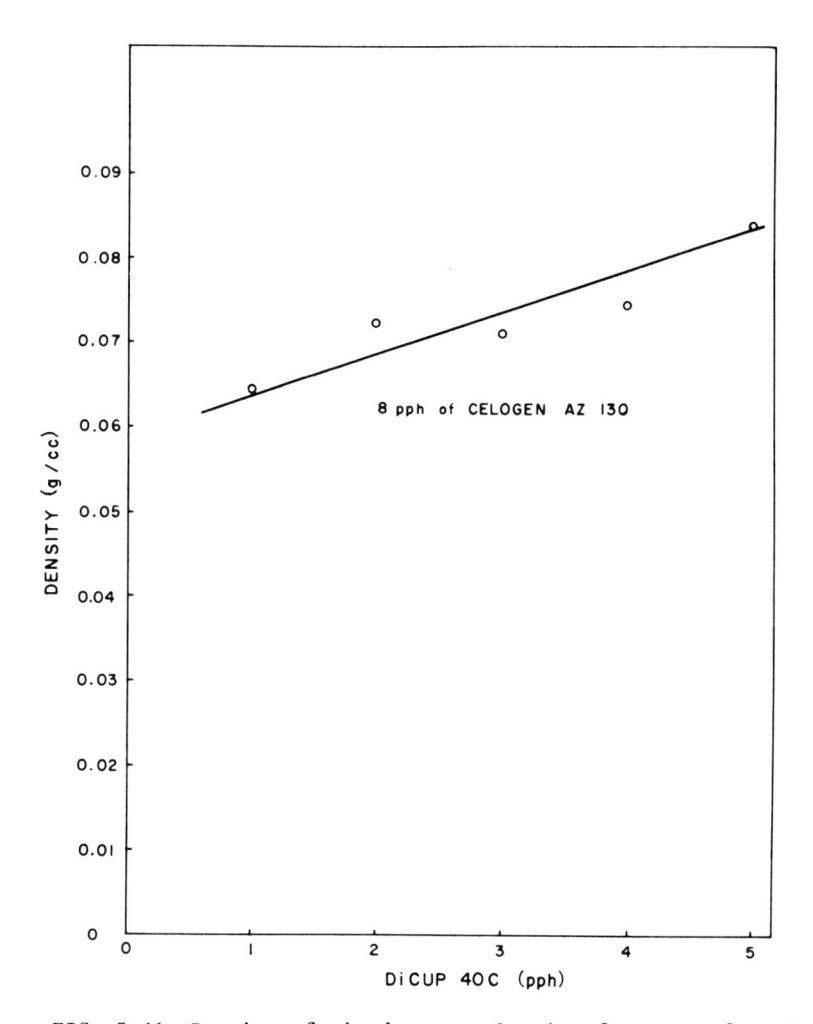

FIG. 5.41 Density of mixed waste plastics foam as a function of degree of cross-linking. (Reprinted from Ref. 31, courtesy Ontario Ministry of the Environment.)

can be achieved with relatively low amounts of peroxide, the cost of raw materials is relatively low. The treatment is only applicable to plastics waste recovered from municipal refuse or to any other mixture of waste plastics containing considerable amounts of PE.

IV. SECONDARY RECYCLING BY COEXTRUSION AND COINJECTION MOLDING

Both coextrusion and coinjection molding allow the production of plastic products having a sandwich structure. The skin material may be a virgin resin characterized by high tensile strength, good scratch resistance, and good color. The core material may include regrind scrap from multilayer regrind, somewhat thermally degraded material, color rejects, and off-grade materials. The choice of core materials is limited by the requirements that they adhere to the outer surfaces and be processible on injection or extrusion machines. The core material may be additionally foamed to reduce the weight of the part. Coextrusion and coinjection molding can be used to produce heavy plastic parts which would otherwise be uneconomical.

A. Coextrusion

Coextrusion is basically an extrusion process utilizing two extruders pumping two different plastics through a single die designed so as to form a product having a sandwich structure (Fig. 5.42). Continuous profiles of sheet, pipe, rod, and other shapes can be produced.

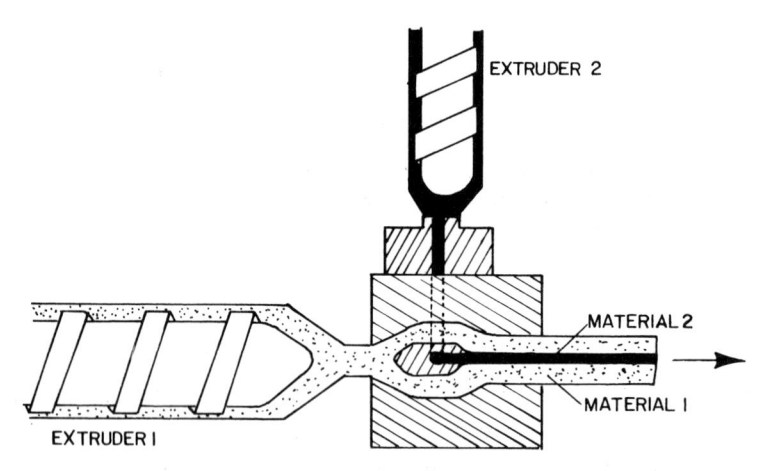

FIG. 5.42 Principle of coextrusion.

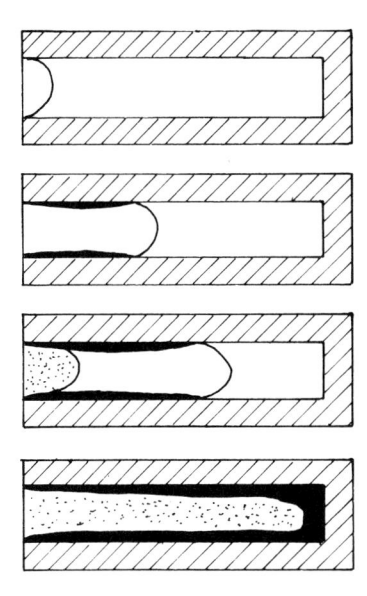

FIG. 5.43 Process sequence in a simple single-channel coinjection.

B. Coinjection Molding

Coinjection molding is technically more complicated than coextrusion. The principles of coinjection molding are shown in Fig. 5.43. Molten plastic enters the mold cavity, where the layer contacting the cold walls solidifies. After the cavity has been partly filled the second material is injected. At that moment a nonmoving layer of solidified plastic exists at the wall with a flowing region in the center of the flow channel. The second plastic melt flows into the center and pushes the soft portion of the first plastic against the walls, creating a sandwich structure [34].

Figure 5.44 shows the three coinjection molding techniques that are used. The one-channel technique involves the sequential injection of two plastic portions into a mold: a specially designed valve allows the injection of material A, and then stops the flow to permit the injection of material B. Due to the flow behavior of the plastic melt and because the skin material cools on the mold surface, a sandwich structure is obtained.

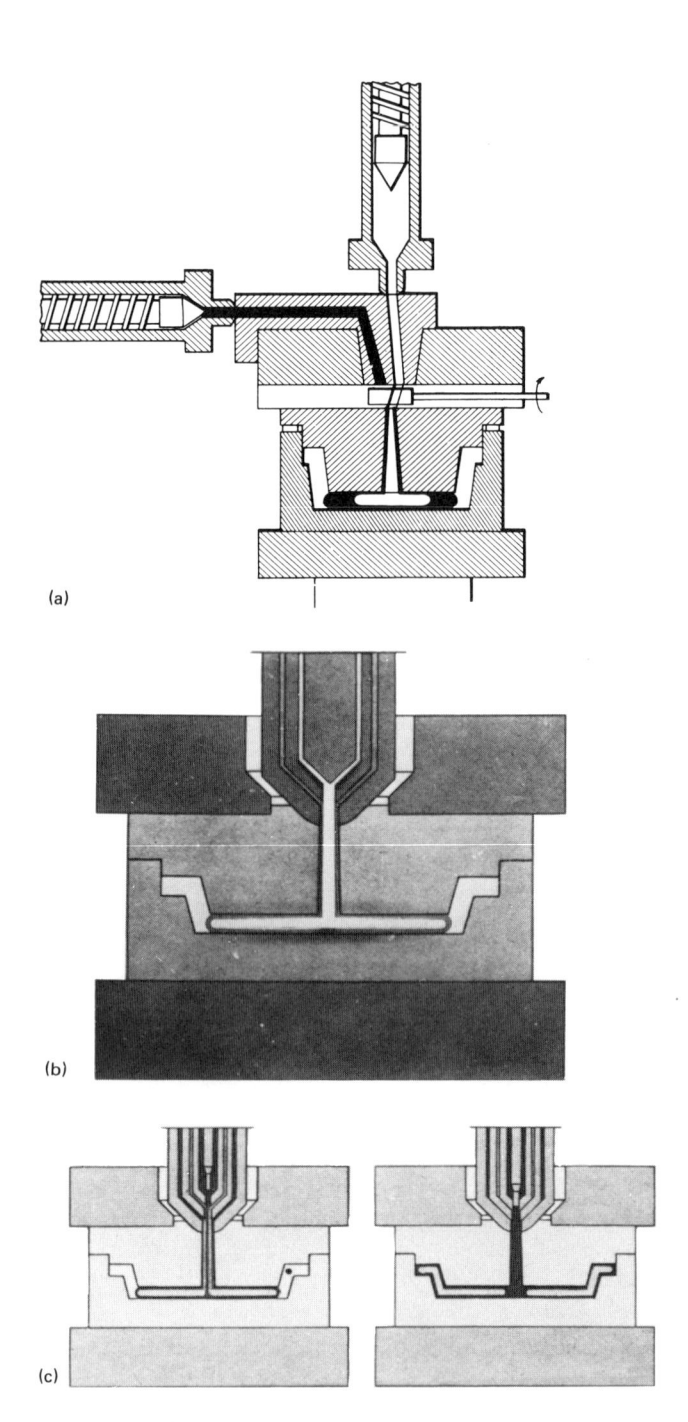

FIG. 5.44 Coinjection molding techniques: (a) one-channel; (b) two-channel; (c) three-channel. (Reprinted from Ref. 35, courtesy the Plastic Institute.)

The two-channel technique enables the two plastic melts to be injected simultaneously: this permits good control of the skin thickness.

The three-channel technique offers a particular advantage over the two-channel technique when the article is of a form allowing central injection. However, when using multicavity and three-plate molds, the two-channel technique is preferred because the nozzle is less complex and the material and color can be changed in a very short time [36].

Coinjection molding is reported to have been used in molding ABS telephone components using regrind from damaged or obsolete telephones as a core, and for flowerpots using waste plastics generated by the packaging industry (coextruded sheet) as a core and PS as a facing component [35-36]. As the equipment for co-injection molding becomes more widely used for molding virgin resins, its use in secondary recycling will also increase.

V. THE USE OF WASTE PLASTICS AS FILLERS

Some waste plastics can be used as fillers/extenders in the same type of plastic, different plastic, or nonplastic materials. Presently, only the reuse of thermosets as fillers for the same or similar thermosets is of practical importance.

Figure 5.45 schematically shows a phenolic sample reinforced with fibrous filler (a). The addition of small amounts of cured reground phenolic (b) will not affect the mechanical properties of the material. If the adhesion between the virgin and the recycled material is good, and since the stiffness of both phases is the same, the filler acts only as an extender without creating stress concentration and weakening the sample. Some slight change in strength might occur due to the discontinuity in the positioning of fibers (no fibers crossing the interfaces). Further addition of regrind (c) creates random spots where the filler particles touch one another. Since very little virgin resin is present

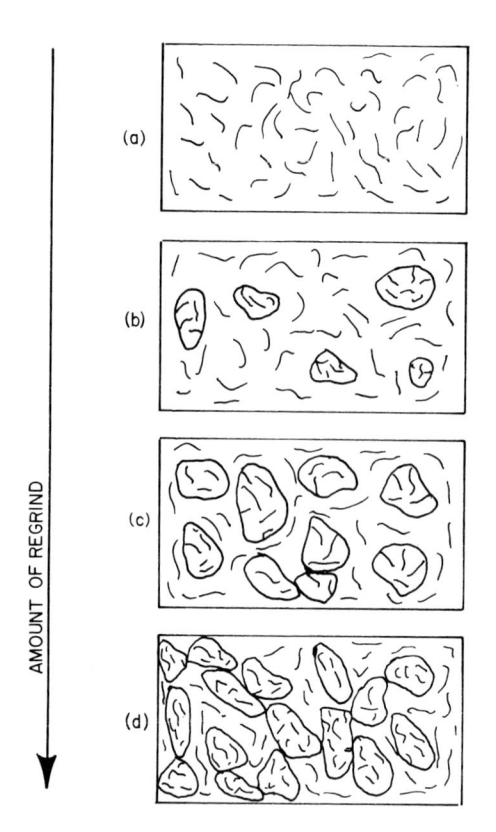

FIG. 5.45 Thermoset molding compound containing regrind.

there, the bonding is poor, and the spots become weak points in the sample. As the number of these weak points increases (d), the strength of the sample drastically decreases. Besides directly affecting the mechanical properties of the thermosetting molding compound, regrind also affects the flow properties and the quality of the product. This secondary effect makes the relationship between the amount of regrind and the mechanical properties of the product somewhat different from what might have been expected from simple analysis. Figures 5.46 and 5.47 show the dependence of the tensile modulus and tensile strength on the mesh size of the regrind for phenolic samples containing 20, 40, and 60% regrind. It

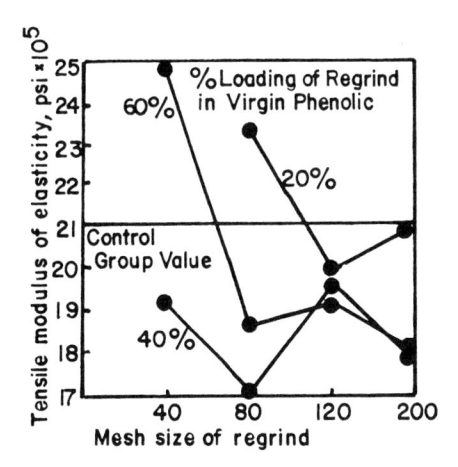

FIG. 5.46 Relationship of mesh size to the modulus of elasticity. (Reprinted from Ref. 37, courtesy Hartman Communications, Inc.)

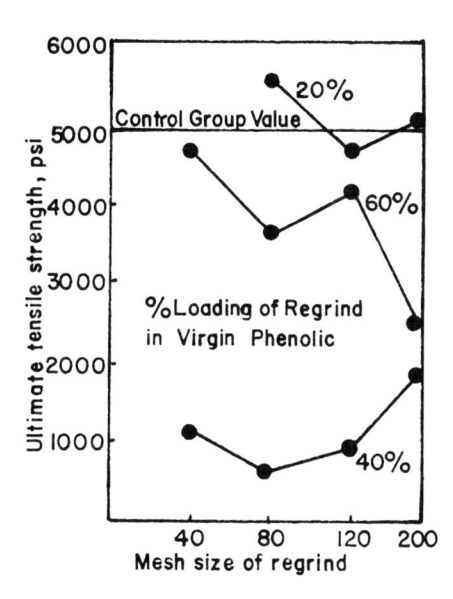

FIG. 5.47 Relationship of mesh size to ultimate tensile strength. (Reprinted from Ref. 37, courtesy Hartman Communications, Inc.)

FIG. 5.48 Hull Recycler--an automated machine for recycling thermoset plastics waste.

is interesting to note that minimum tensile strength and modulus are achieved with 40% regrind: both 20 and 60% regrind give better properties. An 80-mesh regrind at 20% loading increases the strength and elongation of phenolic compounds above the values

Secondary Recycling

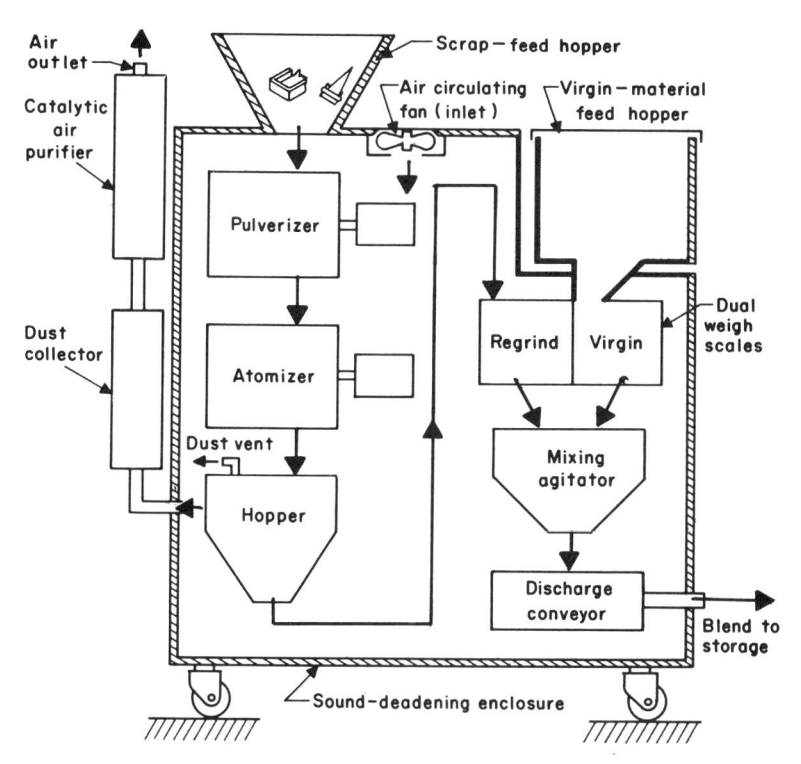

FIG. 5.49 Schematic flow diagram of Hull Recycler. (Reprinted from Ref. 39, courtesy McGraw-Hill.)

for the original compound. Typical properties of phenolic resin filled with up to 15% regrind are shown in Table 5.7.

Although the thermoset plastics waste can be granulated on a number of commercially available granulators and then added to the virgin material, specialized automated equipment for pulverizing and blending is commercially available. Figure 5.48 shows a picture of such a machine manufactured by the Hull Corporation. The operating principles of the Hull Recycler are shown in Fig. 5.49. First the plastics waste is fed into the machine hopper by an operator or conveyor system. From there the material is fed into a series of hammermills and pulverizers which reduce

TABLE 5.7 Typical Properties of Phenolic Molding Compound Filled with up to 15% Regrind

Property	General-purpose phenolic				Medium-heat-resistant phenolic			
	Virgin	Regrind			Virgin	Regrind		
		5%	10%	15%		5%	10%	15%
Specific gravity	1.37	1.38	1.39	1.40	1.58	1.56	1.56	1.56
Moisture retension	5.6%	5.6%	5.6%	5.3%	2.3%	1.4%	2.0%	1.8%
Flexural strength (psi)	12,700	12,300	11,800	11,500	13,400	11,200	10,500	6,000
Impact strength (psi) (Izod, ft-lb/in. notch)	0.35	0.36	0.37	0.35	0.35	0.33	0.34	0.32
Tensile strength (psi)	7,900	7,600	7,700	7,400	8,000	7,300	6,800	5,900
Compressive strength (psi)	32,300	32,600	32,600	32,300	30,600	31,400	31,700	32,000
Flexural modulus (psi)	1.1×10^6	1.1×10^6	1.1×10^6	1.1×10^6	1.4×10^6	1.3×10^6	1.3×10^6	1.1×10^6

Reprinted from Ref. 40, courtesy Society of Plastics Engineers, Inc.

the waste to 20-mesh size. In the second series of hammermills the material is further reduced to 200 mesh. The reground material is transported internally within the machine to a set of weighscale feeders, and thoroughly blended with virgin materials in a blender. After blending, the material is discharged through a set of internal conveyers to the outlet of the machine, where it is collected in storage drums. Since the pulverizing operation can raise the temperature of the materials, causing them to emit gases similar to those emitted during the postcuring operation, the machine has been equipped with a catalytic converter for turning noxious fumes into water and nontoxic gases. Thermosetting molding compounds such as ureas, melamines, dially phthalates, epoxies, polyesters, and alkyds can be recycled by means of the Hull Recycler [38,40].

REFERENCES

1. J. N. Patterson, "The Effect of Dispersion on the Mechanical Properties of Polymer Blends," *35th ANTEC SPE*, Montreal, 1977.
2. D. R. Paul, C. E. Vinson, and C. E. Locke, "The Potential for Reuse of Plastics from Solid Wastes," *Polym. Eng. Sci. 12(3)*, 1972.
3. H. V. Smith, "Some Criteria for the Successful Commercial Recycling of Heterogeneous Plastics Waste," *Recycling World Congress*, Basel, 1978.
4. "Japanese Process Makes Recycling Profitable," *Eur. Plast. News, 1(7)*, 1974.
5. "Reverzer," Technical Brochure, Mitsubishi, Petrochemical Co. Ltd.
6. "Recycling Equipment for Mixed Scrap," *Polym. Age 5(12)*, 1974.
7. M. D. Lever, "The Reprocessing of Thermoplastic Waste into a New Material of Fabrication," *Conf. Publ. Waste Process Ind.*, Institution of Mechanical Engineers, 1977.
8. M. D. Lever and R. S. Parker, "Manufacture and Use of Shapes, Sheets and Assembled Products from Once-used Classified Thermoplastics," *Plast. Rubber Proc. 3(3)*, 1978.
9. "Advanced Concept in Reuse of Plastics Waste," Technical Brochure, Mitsubishi Petrochemical Co. Ltd.

10. "The Reverzer," *REHSIF Bulletin*, No. 300.
11. "The Klobbie," *REHSIF Bulletin*, No. 400, Rehsif S. A., Switzerland.
12. H. V. Smith, "The Recycling of Mixed Thermoplastics Waste," *Polym. Plast. Technol. Eng. 12(2)*, 1979.
13. H. Frederix, "Recycling of Mixed Plastics Wastes Through Plastification," *Kunststoffe 68(5)*, 1978.
14. W. A. Mach, *Recycling Plastics: The Problems and Potential Solutions,* ASTM Special Technical Publication 533, Philadelphia.
15. W. Mack, "Recycling Plastics at a Profit," *28th Annual Western Conference of SPI*, May 1971.
16. "Waste Plastic Regeneration by the Regal Converter," Technical Brochure, Plastic Recycling Ltd.
17. "Recycling for Long Term End Use," *Eur. Plast. News 1(80)*, 1974.
18. M. R. Paske, "Regal Process for Plastics Waste," *Polym. Age 5(9)*, 1974.
19. "Dump or Recycle," *Polym. Age 4(9)*, 1973.
20. "Tufbord," Technical Brochure, Reclamat Int. Ltd.
21. "K Board Will Suit You Too!" Advertising Brochure, Kabor Ltd.
22. "The Recycling Dream Is Turning into Reality," *Mod. Plast. 49(9)*, 1972.
23. J. N. Schramm and R. R. Blanchard, "The Use of CPE as a *Compatibilizer* for Reclamation of Waste Plastic Materials," Presented at *Palisades Section SPE RETEC, Plastics and Ecology*, Cherry Hill, N.J., Oct. 1970.
24. D. R. Paul, C. E. Locke, and C. E. Vinson, "Chlorinated Polyethylene Modification of Blends Derived from Waste Plastics, Part I: Mechanical Behavior," *Polym. Eng. Sci. 13(3)*, 1973.
25. C. E. Locke and D. R. Paul, "Chlorinated Polyethylene Modification of Blends Derived from Waste Plastics Part II: Mechanism of Modification," *Polym. Eng. Sci. 13(4)*, 1973.
26. A. Ram, M. Markis, and J. Kost, "Reuse of Plastics from Solid Wastes," *Polym. Eng. Sci. 17(4)*, 1977.
27. C. E. Locke and D. R. Paul, "Graft Copolymer Modification of Polyethylene-Polystyrene Blends. I. Graft Preparation and Characterization," *J. Appl. Polym. Sci. 17(8)*, 1973.
28. C. E. Locke and D. R. Paul, "Graft Copolymer Modification of Polyethylene-Polystyrene Blends. II, Properties of Modified Blends," *J. Appl. Polym. Sci. 17(9)*, 1973.

29. W. M. Barentsen and D. Heikens, "Mechanical Properties of Polystyrene/Low Density Polyethylene Blends," *Polymer 14(10)*, 1973.

30. W. M. Barentsen and D. Heikens, "Effect of Addition of Graft Copolymer on the Microstructure and Impact Strength of PS/LDPE Blends," *Polymer 15(2)*, 1974.

31. J. Leidner, *Utilization of Recycled Post-Consumer Plastics,* Report for the Ontario Ministry of the Environment, April 1976.

32. J. Leidner, "Recycling of Post-Consumer Plastics," *35th ANTEC SPE*, Montreal, 1977.

33. J. Leidner, "Recovery of the Value from Post-Consumer Plastics Waste," *Polym. Plast. Technol. Eng. 10(2)*, 1978.

34. R. C. Donovan, K. S. Rabe, W. K. Mammel, and H. A. Lord, "Recycling Plastic by Two-Shot Molding," *33rd ANTEC SPE*, May 1975.

35. H. Eckardt and S. Davies, "An Introduction to Multicomponent (Sandwich) Injection Moulding," *Plast. Rubber Int. 4(2)*, 1979.

36. W. Fillman, "Use of Recycled Materials and Fillers for Making Sandwich Moulding," *Plast. Rubbers Proc. 3(3)*, 1978.

37. R. A. Kruppa, "Why Not Reclaim Phenolic? *Plast. Technol. 23(5)*, 1977.

38. R. A. Kruppa, *Selected Physical Characteristics of Mixtures of Reground and Virgin Phenolic Molding Compound,* Final Report, Faculty Research Committee, Bowling Green State University

39. S. H. Bauer, "Processing Plastics—Recycle Scrap Effectively," *Plast. Eng. 33(3)*, 1977.

40. "Thermoset Injection Molding Becomes More Reliable and More Economical," *Mod. Plast. 56(6)*, 1979.

Chapter 6

TERTIARY RECYCLING: CHEMICALS FROM PLASTICS WASTE

I. PYROLYSIS

A. Pyrolysis of Postconsumer Waste

1. *Introduction to Pyrolysis of Municipal Refuse*

Pyrolysis is defined as the physical and chemical decomposition of organic materials caused by heating in an oxygen-free or oxygen-deficient atmosphere. Pyrolysis is not a new process: it has been used for many years in the manufacture of such wood derivatives as charcoal, methanol, acetic acid, and turpentine. The pyrolysis process is capable of producing simple chemical compounds out of mixtures of waste materials which would otherwise have to be incinerated or disposed of by landfilling. The products of pyrolysis can be employed as commercially useful chemicals or as fuel.

The following advantages are claimed for pyrolysis [1,2]:

1. Most municipal solid waste can be converted into an economically viable form.
2. The volume of waste can be reduced by 90% or more.
3. The pyrolysis process is contained and thus does not cause air pollution.
4. Because the process is nonpolluting and requires little space, pyrolysis plants can be located in cities, resulting in lower transportation costs.
5. The process is a net energy producer.

6. The energy produced is in a convenient form, i.e., gas, oil, and char.
7. The process can be set up so that any valuable chemicals can be recovered.
8. Since little oxidation takes place during the process, metallic components can be recovered after the waste has been pyrolyzed.

Pyrolysis, unlike incineration, is an endothermic reaction and heat must be applied to distill off volatile components. Lamp [3] lists the following reactions taking place during the pyrolysis of solid waste:

Principal reaction:

Organic material $\rightarrow$ gases + liquid + char

Secondary reactions:

$$CO + H_2O \rightleftharpoons CO_2 + H_2 + \text{heat}$$
$$C + H_2O \rightleftharpoons CO + H_2 - \text{heat}$$
$$C + CO_2 \rightleftharpoons 2CO - \text{heat} \tag{1}$$
$$C + O_2 \rightleftharpoons CO_2 + \text{heat}$$
$$C + 2H_2 \rightleftharpoons CH_4 + \text{heat}$$

As shown, the products of the pyrolysis of solid waste are in the form of gases, liquids, and char. The U.S. Bureau of Mines [4] reported the results of an investigation of the pyrolysis of both raw municipal and processed municipal wastes (Table 6.1). The raw municipal waste used was a typical municipal refuse containing about 1.5% plastics, while the processed waste contained mainly municipal plastics waste.

The solid residue from the process was in the form of a lightweight, flaky char, which could be coarsely sieved to remove extraneous materials such as bottle caps and lids. The solid residue obtained from the processed municipal refuse (waste plastics) had a higher heating value than that obtained from the raw municipal waste. The sulfur content was below 0.4%. Tests

TABLE 6.1 Yields of Products from Pyrolysis of Municipal Refuse

Refuse	Pyrolysis temp. °C	Yield wt % of refuse						Yield per ton of refuse				
		Residue	Gas	Tar	Light oil in gas	Free ammonia	Liquor	Total	Gas (ft³)	Tar, (gal)	Light oil in gas (gal)	Liquor (gal)	Ammonium sulfate (lb)
Raw municipal	500-900	9.3	26.7	2.2	0.9	0.05	55.8	94.5	11,363	4.8	1.5	133.4	17.9
	750	11.5	23.7	1.2	0.9	0.03	55.0	92.3	9,620	2.6	2.5	131.6	23.7
	900	7.7	39.5	0.2	—	0.03	47.8	95.2	17,741	0.5	—	113.9	25.1
Processed municipal containing plastic film	500-900	21.2	27.7	2.3	1.3	0.05	40.6	93.2	11,545	5.6	3.7	96.7	18.2
	750	19.3	18.3	1.0	0.9	0.02	51.5	91.2	7,580	2.2	2.6	122.6	28.4
	900	19.1	40.1	0.6	0.2	0.04	35.5	95.3	18,058	1.4	0.6	97.4	31.5

Reprinted from Ref. 4, courtesy Bureau of Mines, U.S. Department of the Interior.

have shown that the char could easily be briquetted with suitable cinders, producing a solid fuel.

The major constituents of the gas fraction in the U.S. Bureau of Mines experiments were hydrogen, carbon monoxide, methane, and ethylene (Table 6.2). Pyrolysis at lower temperatures yielded gas with a higher heating value, and pyrolysis at higher temperature, because of its higher gas yield, resulted in a gaseous fraction with a higher total energy content per ton of refuse. The heat value of the gaseous fraction can be upgraded by the removal of carbon dioxide. Since approximately 2 million Btu are required to pyrolyze 1 ton of municipal refuse at 900°C and the process produces gas with a calorific value at 8 million Btu, there is more than enough gas to sustain the process [4].

The liquid portion of the pyrolysis products consists of tar, light oils, and liquor. The tar component of the pyrolysis products obtained in the U.S. Bureau of Mines experiments is characterized in Table 6.3, and the analysis of light oils is given in Table 6.4.

The liquor obtained in the pyrolysis of municipal refuse is 94 to 100% water containing small amounts of organic and inorganic compounds (Table 6.5). These compounds, because of their low concentration levels, have no commercial value; together with the water they may present a significant waste disposal problem.

2. Parameters Influencing the Yields of Pyrolysis Processes

Yields of specific products of a pyrolysis process are influenced by residence time, temperature, particle size of waste feed, and atmosphere (oxygen, air, oxygen-free, steam). The effects of these parameters on the yields of the pyrolysis process are shown in Fig. 6.1.

Since the pyrolysis process is endothermic, heat has to be applied to the waste being pyrolyzed. There are two ways in which heating can be accomplished: (1) by partial combustion of the waste and/or supplementary fuel (direct method); and (2) by

TABLE 6.2 Analysis of Gases from Pyrolysis of Municipal Waste

Pyrolysis temperature (°C):	Raw municipal refuse			Processed municipal refuse containing plastic film		
	500-900	750	900	500-900	750	900
Analysis (vol %)						
Hydrogen	45.47	30.86	51.91	44.86	25.27	42.41
Carbon monoxide	21.54	15.57	18.16	19.62	25.09	20.16
Methane	13.15	22.57	12.66	18.73	17.57	13.92
Ethane	1.30	2.05	0.14	2.08	2.01	0.25
Ethylene	4.67	7.56	4.68	4.54	10.36	7.89
Carbon dioxide	11.41	18.44	11.42	8.02	18.25	13.91
Propane	<0.01[a]	<0.01	<0.01	<0.01	<0.01	1.17
Isobutane	Trace[b]	Trace	Trace	Trace	Trace	Trace
Butane	0.08	0.01	0.44	0.03	Trace	0.11
Butene-1	0.16	0.15	Trace	0.11	Trace	Trace
Isobutylene	0.17	0.15	Trace	0.15	Trace	Trace
trans-Butene-2	0.04	0.03	Trace	0.08	Trace	Trace
cis-Butene-2	0.07	Trace	Trace	0.03	Trace	Trace
Pentane	Trace	Trace	Trace	0.16	Trace	Trace
Pentene	0.60	0.87	0.20	0.17	0.53	0.07
Unidentified	<0.01	0.21	0.06	0.06	0.15	0.01
Btu/ft^3 of gas	473	563	447	536	570	511
Million Btu/ton of refuse pyrolyzed	5473	5421	7930	6188	4207	9228

[a] <0.01 = 1 part in 10^4.
[b] Trace = less than 1 part in 10^5.
Reprinted from Ref. 4, courtesy Bureau of Mines, U.S. Department of the Interior.

separating the combustion chamber from the pyrolysis chamber, using a heat exchange medium to transfer the heat (indirect method).

In the direct method oxygen or air has to be supplied to the pyrolysis reactor. Oxygen is more expensive, but the process yields gas which has a much higher Btu content, and the potential

TABLE 6.3 Yields of Tar Components from Pyrolysis of Municipal Refuse at 500-900°C

(Gallons per ton of refuse)

Refuse	Tar Acids	Bases	Neutral oil	Residue	Olefins	Aromatics	Paraffins and naphthenes	Pounds per ton of refuse Anthracenes	Naphthalenes
Raw municipal	4.8 0.4	0.2	1.5	2.7	0.3	1.1	0.2	—	—
Processed municipal containing plastic film	5.6 0.3	0.2	1.5	3.5	0.4	1.0	0.2	—	—
Bell mill industrial	4.1 0.2	0.2	0.8	2.9	0.2	0.6	0.1	—	—
Condard mill industrial	1.7 0.1	0.1	0.5	1.0	0.1	0.3	0.1	—	—

Reprinted from Ref. 4, courtesy Bureau of Mines, U.S. Department of the Interior.

TABLE 6.4 Chromatographic Analysis of Light Oils from Pyrolysis of Municipal Refuse

	Raw municipal refuse, pyrolysis temp. (°C)			Processed municipal refuse containing plastic film, pyrolysis temp. (°C)		
Substance	500-900	750	900	500-900	750	900
Prebenzene	25.04	3.70	0.80	28.08	0.74	0.79
Benzene	37.54	78.47	73.39	57.35	85.18	92.10
Toluene	23.76	14.06	12.25	10.31	11.54	3.82
Ethylbenzene	2.50	0.31	0.02	1.38	0.14	<0.01
m,p-Xylene	2.39	0.65	2.84	0.71	0.47	0.67
o-Xylene	1.01	0.20	0.81	0.36	0.20	0.16
Unidentified	7.76	2.61	9.89	1.81	1.73	2.45

Reprinted from Ref. 4, courtesy Bureau of Mines, U.S. Department of the Interior.

environmental problems caused by the formation of NO_x compounds are avoided. The steam can be used to reduce the amount of solid residue (by water-carbon reaction) and to shift the carbon monoxide/hydrogen ratio.

TABLE 6.5 Water Content of Liquor Produced in the Pyrolysis of Municipal Refuse

		% Water content	
Refuse	Pyrolysis temp. (°C)	Liquor from condensors	Liquor from tar trap
Raw municipal	500-900	96.1	96.2
	750	98.2	94.5
	900	98.9	94.6
Processed municipal containing plastic film	500-900	97.4	99.7
	750	100.0	98.9
	900	99.3	91.5

Reprinted from Ref. 4, courtesy Bureau of Mines, U.S. Department of the Interior.

Tertiary Recycling

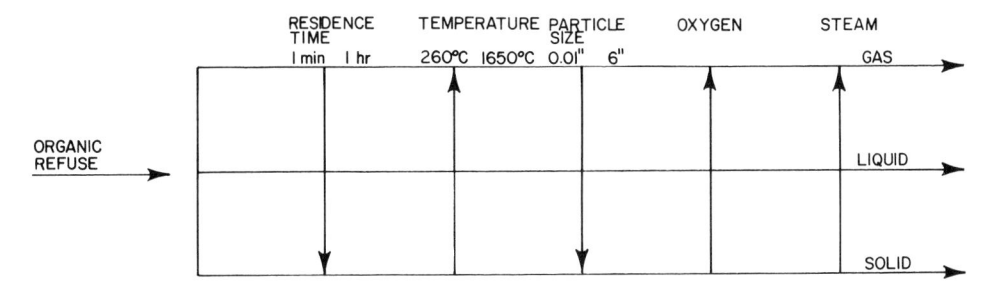

```
                RESIDENCE  TEMPERATURE PARTICLE   OXYGEN      STEAM
                TIME                   SIZE
                1 min  1 hr 260°C 1650°C 0.01"  6"                      GAS →

ORGANIC
REFUSE →                                                                LIQUID →

                                                                        SOLID →
```

FIG. 6.1 Parameters influencing yields of pyrolysis process.

An increase in pyrolysis temperature has the following effect on product yield:

1. The solid residue decreases with increasing temperature due to an increased conversion of carbon into gas products.
2. The amount of water decreases with the increasing temperature due to reactions with methane and carbon monoxide:

$$CH_4 + H_2O \; (g) \rightarrow CO + 3H_2$$
$$CO + H_2O \; (g) \rightarrow CO_2 + H_2$$
(2)

3. The amount of light oils condensed decreases with increased temperature due to cracking reactions which produce products with lower molecular weight.
4. Gas yields and heating values increase with temperature [6].

Feedstock preparation is another important parameter influencing yields. If the feed is prepared by shredding it into smaller and uniform particles, the process operator is more capable of controlling such parameters as temperature, residence time, and heating rates. For a given reactor temperature and residence time smaller particle size will result in faster reaction rates, reduction in the amounts of solid and liquid products, and an increase in gas generation.

Predrying of feedstock is a common practice. The use of predried feedstock reduces the amount of heat input to the reactor, the heating time, and the amount of liquid generated. These advantages are offset by the additional step and the equipment required.

3. Types of Pyrolysis Systems

Most pyrolysis systems are very similar. Before municipal solid waste is charged into a pyrolysis reactor, inert materials such as metals, glass, and soil may be removed and the feedstock may be predried and shredded. The prepared feedstock is then charged into the pyrolysis reactor. Oxygen or air (in the case of direct heating reactors) or heat exchange medium (in the case of indirect heating reactors) is then introduced to the reactor. The heat contained in the pyrolysis products is recovered and the products possibly upgraded by final processing.

A number of reactor types have been used for pyrolysis, the most popular being the shaft reactor, the rotary kiln reactor, and the fluidized bed reactor [5]. Shaft reactors, the simplest and least expensive type, can be designed to operate vertically or horizontally. In the vertical reactors (Fig. 6.2), municipal solid waste is fed from the top and settles under its own weight toward the bottom. Oxygen, air, or heat exchange medium is fed at the bottom of the reactor. The gases generated in the process pass upward and are removed from the top. In the horizontal shaft reactors a feed conveyer system is used to move the feedstock through the reactor. Although the feed and the removal of slag are simplified in these reactors, the reliability of the conveyer at elevated temperature can be a problem [5].

A schematic of a rotary kiln reactor is shown in Fig. 6.3. The reactor is a cylinder (usual length-to-diameter ratio 4:1 and 10:1) rotated on suitable bearings and slightly inclined. Feed is introduced at one end of the reactor and progresses through it due to the rotation and slope. Mechanisms for feeding and removal of the waste are provided. The rotary kiln reactor is mechanically more complex than the shaft reactor but provides the advantage of mixing the feed during pyrolysis.

Fluidized bed pyrolysis reactors have been used in coal gasification processes. Those used in the pyrolysis of solid waste (Fig. 6.4) require preshredded feed, since fluidization

Tertiary Recycling

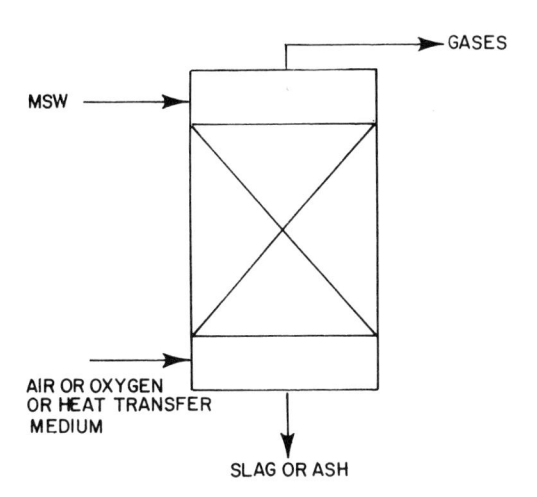

FIG. 6.2 Vertical shaft reactor.

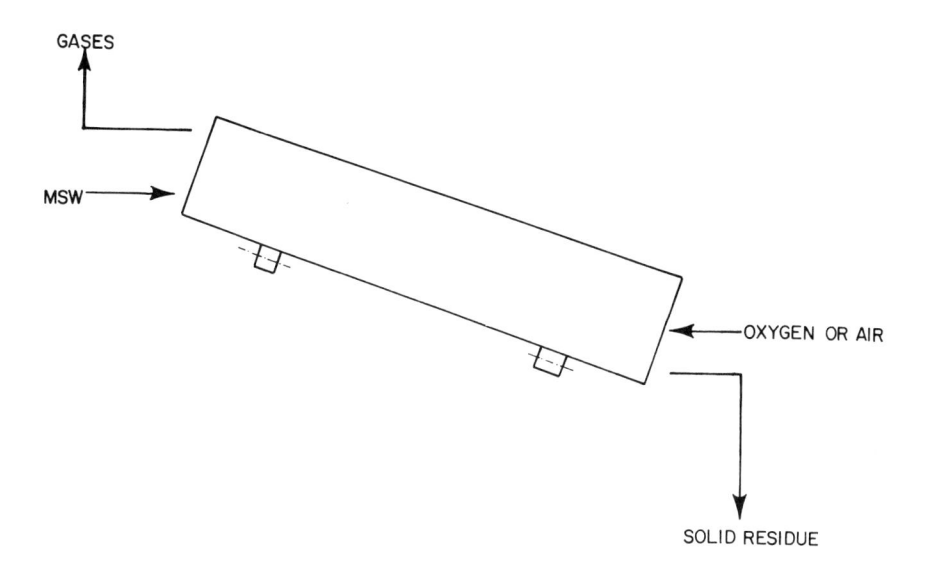

FIG. 6.3 Rotary kiln pyrolysis reactor.

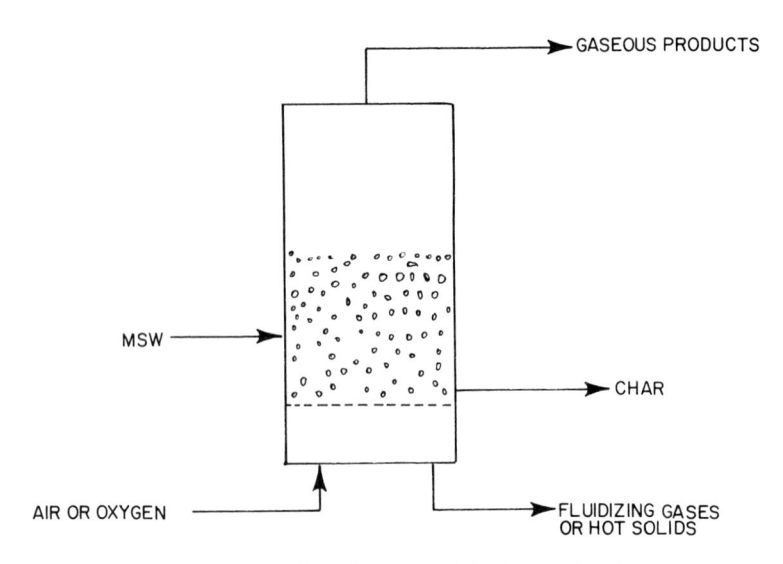

FIG. 6.4 Schematic of a fluidized bed pyrolysis reactor.

requires reasonably uniform material. These reactors operate at lower temperature ranges (1400 to 1800°F), below the temperature at which slag is formed. Heat necessary for the pyrolysis process can be generated by partial oxidation of the waste or by recirculating preheated fluidized solids. In some systems a second fluid bed reactor is used in which material such as sand or limestone is circulated. The char in combination with oxygen or air is used to heat the solid medium, which is then used to transfer heat to the feedstock in the pyrolysis reactor.

The fluidized bed reactor has the disadvantage of requiring considerable feedstock preparation; its main advantage is good temperature control. Table 6.6 summarizes the main operational features of the various types of pyrolysis reactors.

4. *Examples of Solid Waste Pyrolysis Systems*

Table 6.7 summarizes the main features of some commercial, pilot plant, or research solid waste pyrolysis systems. These systems

TABLE 6.6 Reactor Type Characteristics

	Direct heating		Indirect heating			
			Wall transfer		Circ. medium	
	Operational simplicity	High heating rate	Operational simplicity	High heating rate	Operational simplicity	High heating rate
Vertical shaft	+		+	-	-	+
Horizontal shaft	None	None	-	-	-	+
Rotary kiln	+		+	-	-	+
Fluidized bed	-	+	None	None	-	+

Note: A plus (+) entry indicates a virtue while a minus (-) entry indicates a detriment; "None" indicates that no process development has been reported in that category; no entry implies neither a virtue nor a detriment.

Reprinted from Ref. 5, courtesy NASA.

TABLE 6.7 Pyrolysis Reactor Classifications

	Heating method		Product distribution			Feed conditions			Reactor temp. (°C)	Res.	Status		
	Direct	Indirect	Solid (Btu/lb)	Liquid	Gas	Raw	Size	Separation			Pilot PLT (TPD)	Comm. (TPD)	
Vertical shaft													
Garrett		X	9,700	10,500	550	X		X	900		4	200	
URDC	X				150	X			2,600		120		
Torrax	X				150	X			3,000		75		
Union Carbide	X				300	X			3,000		5	200	
Horizontal shaft													
Barber-Colman		X			500		X	X	1,200		1		
Rotary kiln													
Monsanto	X		2,500		130		X	X	1,800		35	1,000	
Devco			X		X		X	X	1,000		120	1,500	
Fluidized bed													
W. Virginia	X				450		X	X	1,400	X			

Reprinted from Ref. 5, courtesy NASA.

Tertiary Recycling 231

are designed to handle municipal solid wastes of which plastics constitute only a small proportion.

   a. *Vertical Shaft Reactors*

   Garret System. Figure 6.5 is a simplified flow diagram of the process developed by the Garret Research and Developmental Company (GRDC), a subsidiary of the Occidental Petroleum Corporation. The feed preparation is an integral part of the Garret process, and consists of: (1) primary shredding to under 3 in. in size, (2) removal of most inorganics by air classification, (3) drying to 3% moisture, (4) reduction of inorganic content to 4% by screening, and (5) secondary shredding. Finely shredded organic refuse is rapidly heated in the reactor to produce fuel oil, gas, and solid residue. One barrel of low-sulfur synthetic fuel oil can be recovered from 1 ton of municipal refuse [6].

   URDC System. Figure 6.6 is a flow diagram of the pyrolysis system developed by the Urban Research and Development Corporation. The solid waste is fed at the top of the vertical shaft and dries as it moves downwards. Organics are pyrolyzed and the char and inorganics fall to the combustion zone located at the bottom. Pyrolysis gas is burnt to preheat combustion air. The combustion of char in preheated air provides heat for the pyrolysis of the waste [5].

   Andco-Torrax System. The Andco-Torrax System is similar in principle to the URDC system. The major difference lying in the preheating of the air. The Andco-Torrax system utilizes natural gas for this purpose, after which the energy is recovered by burning the pyrolysis gas to generate steam [7].

   Union Carbide System. The Union Carbide Linde Division Purox system (Fig. 6.7) accepts solid waste, grinds it, removes metals, and moves nonmagnetic material to the top of a 40-ft-high shaft furnace. The pyrolysis reactor consists of three stages, called the drying, reaction, and firing zones. The material is prepared

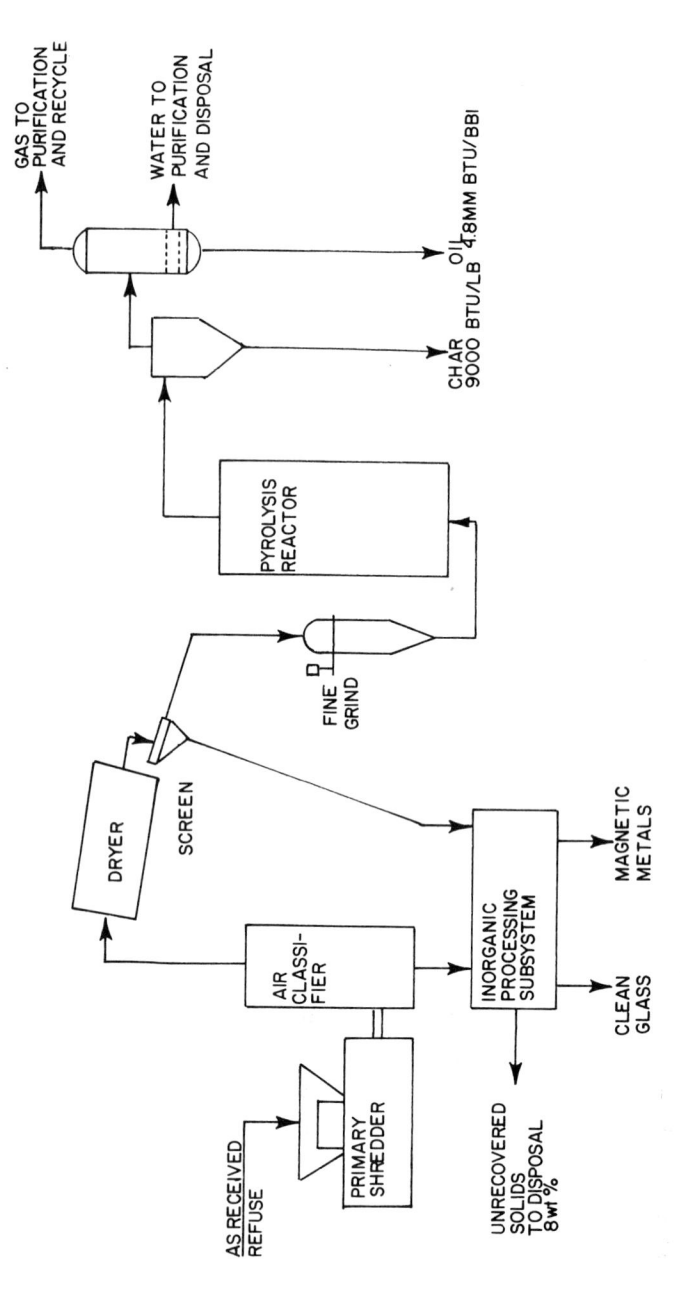

FIG. 6.5 Flow diagram of the Garret process.

Tertiary Recycling   233

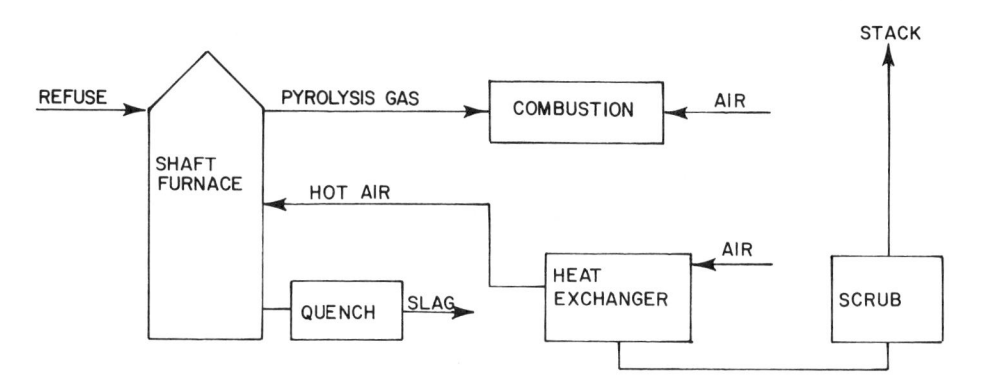

FIG. 6.6 Flow diagram of the URDC pyrolysis process.

in the drying zone before entering the reaction zone. In the reaction zone organics are pyrolyzed, producing combustible gases and a char. The gas is removed, cleaned, and supplied to the user. The char is burned in the firing zone with pure oxygen. The inorganics form slag, while carbon monoxide, from combustion of the char, passes to the reaction zone. The fuel gas produced in the Purox process has a calorific value of 10,600 to 11,200 kJ/m^3 and

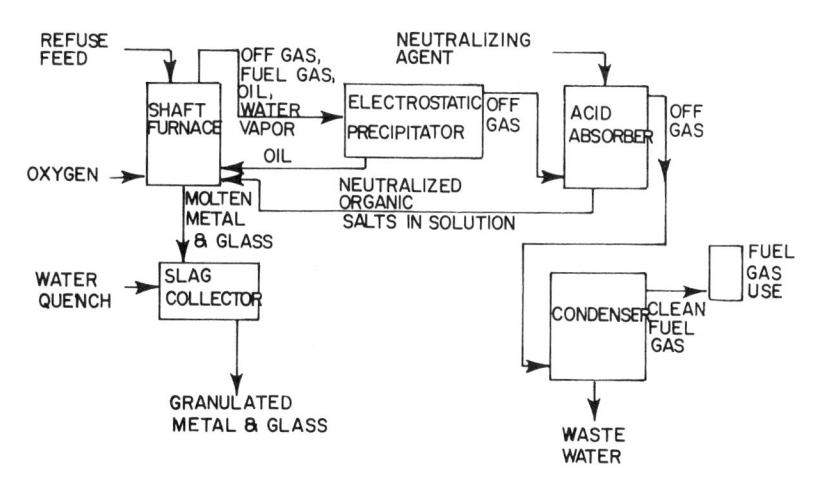

FIG. 6.7 Flow diagram of the Union Carbide process.
(Reprinted from Ref. 5, courtesy NASA.)

can be burned in existing boilers designed for use with natural gas. The usable energy yield of the process is approximately 65% [8].

b. *Horizontal Shaft Reactors*

Barber Colman System. The Barber-Colman pyrolysis system (Fig. 6.8) utilizes a bed of molten lead as the heat transfer medium. After the large metal pieces have been removed from the refuse the feed is shredded and fed to the reactor. Here the feed floats on the surface of the molten lead which is heated from the top by radiant tube burners. The refuse is pyrolyzed at approximately 650°C. Inert materials floating on the surface of the lead are removed by a mechanical rake device. The process generates fuel gas with calorific value of 18,600 to 26,000 kJ/m^3. About 25% of the gas is used in the process; the remaining 75% is available for sale [5].

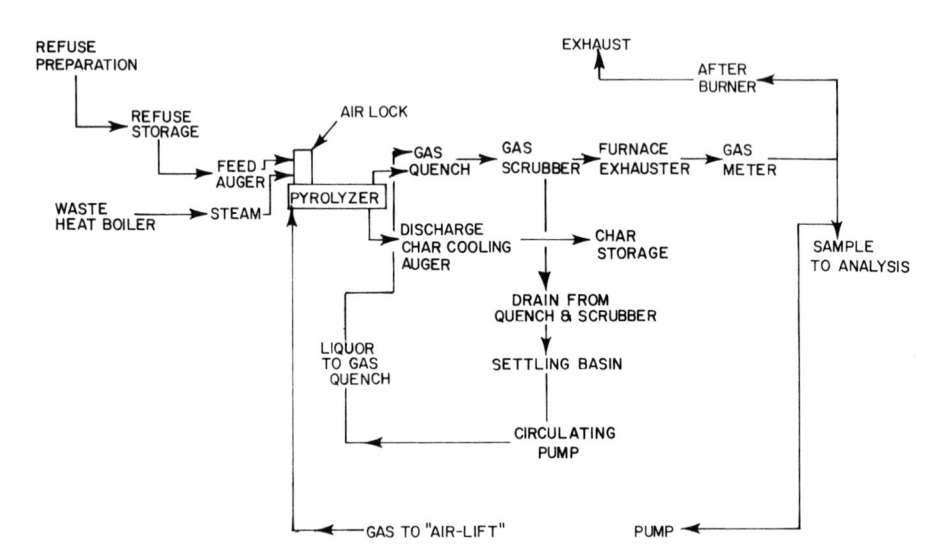

FIG. 6.8 Flow diagram of the Barber-Colman pyrolysis system. (Reprinted from Ref. 5, courtesy NASA.)

c. *Rotary Kiln Reactors*

Monsanto's Landgard System. Figure 6.9 shows a schematic of the Landgard process. Waste received from trucks is transferred to the shredders, and after shredding goes into storage. From there, shredded waste is continuously fed into the rotary kiln. Heat is generated by burning fuel oil. Gases and solids flow countercurrently, which exposes the solids to progressively higher temperatures. The solid residue is discharged from the kiln and transferred to a separate building where it is separated into ferrous metals, glassy materials, and char. Hot gases generated in the process pass through the heat exchangers where steam is generated. This steam can be used for heating and air conditioning.

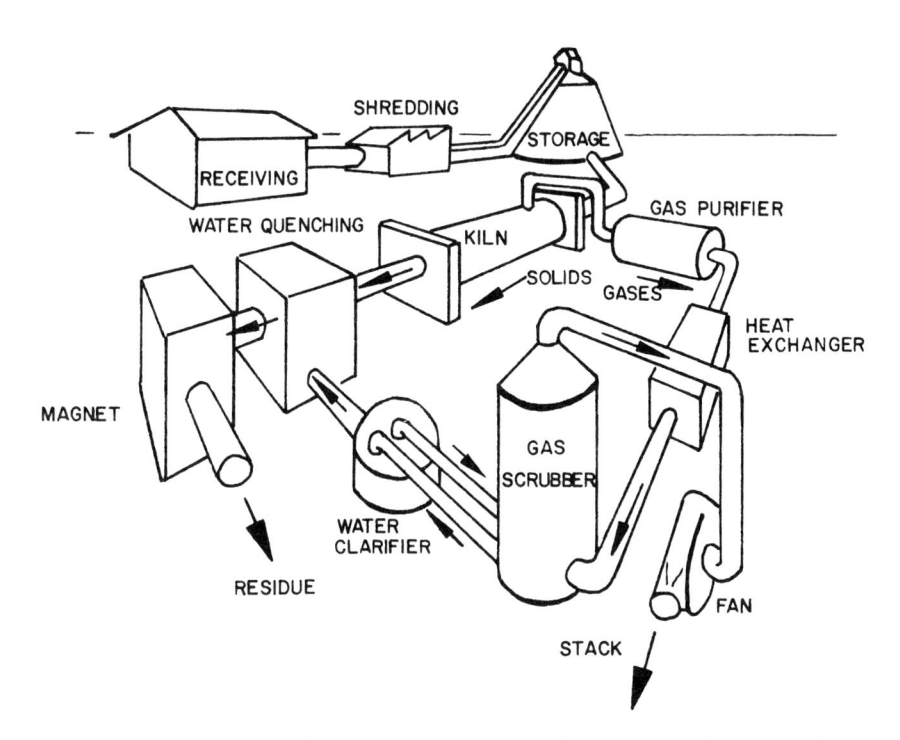

FIG. 6.9 Schematic of Monsanto's Landgard system. (Reprinted with permission from "Disposing of Solid Wastes by Pyrolysis," *Environmental Science and Technology, 9(2)*, 1975. Copyright by the American Chemical Society.)

Exit gases from the boilers are further cleaned and can be burned on site to generate steam or cocombusted with oil, coal, or natural gas by utility, industrial, or institutional boilers [9].

   d. *Fluid Bed Reactors*

   West Virginia University System. Figure 6.10 is a flow diagram of the solid waste pyrolysis system proposed by researchers at West Virginia University. Two fluidized beds are used, one for combustion and the other for pyrolysis. Sand is circulated between these two beds, transferring combustion heat to the pyrolysis bed and char to the combustion bed. No air is present in the pyrolysis bed, resulting in a higher calorific value of the produced gas. A portion of the pyrolysis gas is used as the fluidizing gas in the pyrolysis chamber. Preheated air is used to fluidize the combustion bed and for combustion of the char. The process is expected to produce gas with heating value of 400 to 500 Btu/scf [5].

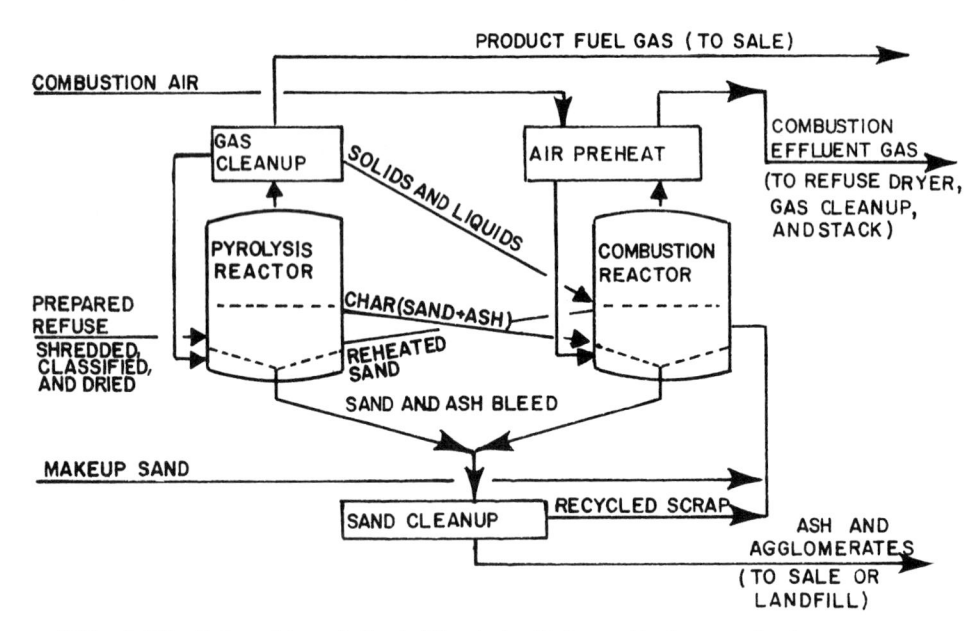

FIG. 6.10 Schematic of West Virginia University's pyrolysis process. (Reprinted from Ref. 5, courtesy NASA.)

B. Pyrolysis of Plastics Waste

The basic concepts of plastics pyrolysis are similar to those involved in the pyrolysis of municipal refuse. The difference lies in the handling characteristics of plastics compared to those of municipal solid waste, and in the type of products obtained in the process. The main commercially useful product recovered in the pyrolysis of municipal solid waste is fuel gas with rather low Btu content, while the pyrolysis products of plastics can be used either as fuel or as feedstock for the chemical industry.

1. PYROLYSIS REACTIONS

The depolymerization of polymers is somewhat similar to the process of polymerization. Polymers decompose into smaller molecules or monomers, depending on their structure and the conditions of the reaction. Polymers that decompose to monomers are shown in Table 6.8. Although the mechanisms of polymerization and depolymerization are similar, the energy required for depolymerization is greater than that for polymerization. If the rate constants of both processes are plotted as a function of inverse absolute temperature, they will cross at a certain temperature. Above that ceiling temperature a net depolymerization will take place.

During the pyrolysis process, the following reactions take place: (1) depolymerization, producing monomers; (2) chain fragmentation, producing low-molecular-weight materials; (3) production of unsaturated compounds; cross-linking of the polymer; and char formation.

The kinetic behavior of polymers during thermal degradation is intermediate between the following extreme possibilities:

1. A large initial production of gaseous products with a very slow decrease in the molecular weight of the polymer (the monomer is being removed from the ends of the chains: a typical example is polymethyl methacrylate).

TABLE 6.8 Thermal Depolymerization Reactions

Polymer	Reaction	Monomer yield
PMMA	$-CH_2-\underset{\underset{COOCH_3}{\|}}{\overset{\overset{CH_3}{\|}}{C}}- \rightarrow CH-\underset{\underset{COOCH_3}{\|}}{\overset{\overset{CH_3}{\|}}{C}}- \rightarrow$ (Stable Nonreactive)	100%
PMS	$-CH_2-\underset{\underset{\phi}{\|}}{\overset{\overset{CH_3}{\|}}{C}}- \rightarrow -\overset{\cdot}{C}H-\underset{\underset{\phi}{\|}}{\overset{\overset{CH_3}{\|}}{C}}- \rightarrow$	100%
PS	$-CH_2-\underset{\underset{\phi}{\|}}{CH}- \rightarrow -CH_2-\underset{\underset{\phi}{\|}}{\overset{\cdot}{C}}- \rightarrow$ (Unstable Reactive)	40% at 300–400°C, more monomer with higher nitrogen pressure. Fragments of <500°C, more fragments with higher nitrogen pressure
PE	$-CH_2-CH_2- \rightarrow -CH_2-\overset{\cdot}{C}H- \rightarrow$	Fragments at 400°C
TFE	$-CF_2-CF_2- \rightarrow -CF_2-\overset{\cdot}{C}F- \rightarrow$	<95% at 500° 16% at 600°C, atmospheric Fragments at 1200°C, low pressure

Reprinted from Ref. 10, courtesy Ann Arbor Science Publishers.

2. A very slow initial production of gaseous products with a rapid decrease in the molecular weight of the polymer (the polymer chain is broken in random locations: a typical example is polyethylene)

The whole range of kinetic behavior between these extremes is possible.

Depending on the design of the pyrolysis process, the resulting product can be in the form of a monomer, a low-molecular-weight polymer, or a mixture of various hydrocarbons.

Tertiary Recycling   239

The TRW Systems Group conducted an analysis of plastics pyrolysis using computer models and the necessary thermodynamic data [11]. Figure 6.11 shows the major products of the thermal decomposition of PE, PS, and PVC.

Initially the decomposition of PS occurs randomly: weak links present in the polymer break first. During this period inhibitors give rise to an induction period. At the end of the induction period, depolymerization takes place. Figures 6.12 to 6.14 show the kinetics of styrene monomer formation at various temperatures, and Figure 6.15 gives feasible reaction path species. The results indicate that benzene is in equilibrium with styrene at 1073 K and 1 atm pressure. Since the formation of that mixture represents the minimum free energy of the system, it is the most probable distribution of species at equilibrium.

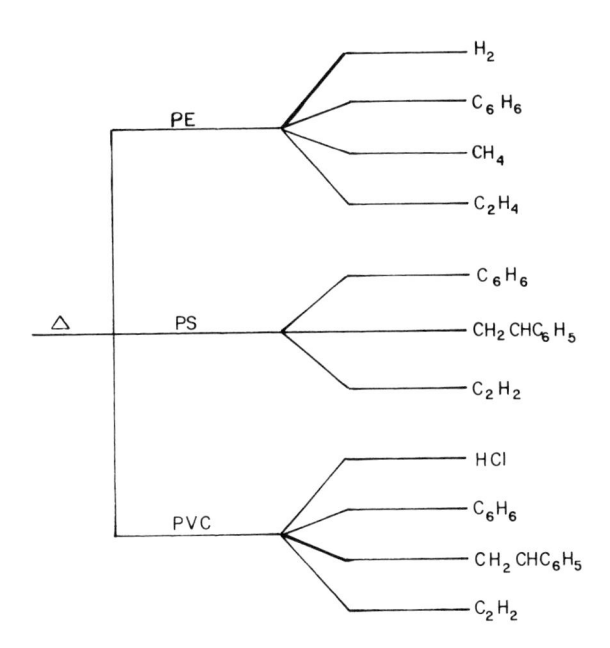

```
              ┌── H₂
         PE ──┼── C₆H₆
              ├── CH₄
              └── C₂H₄

              ┌── C₆H₆
 Δ ──── PS ──┼── CH₂CHC₆H₅
              └── C₂H₂

              ┌── HCl
              ├── C₆H₆
        PVC ──┼── CH₂CHC₆H₅
              └── C₂H₂
```

FIG. 6.11 Major products of thermal decomposition of PE, PS, and PVC. (Reprinted from Ref. 12.)

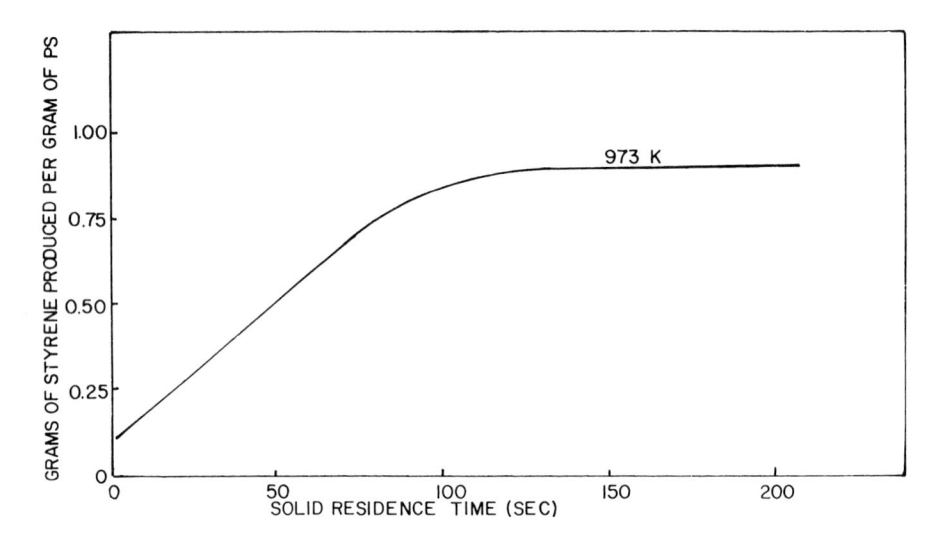

FIG. 6.12 Kinetics of styrene formation of 973 K. (Reprinted from Ref. 12.)

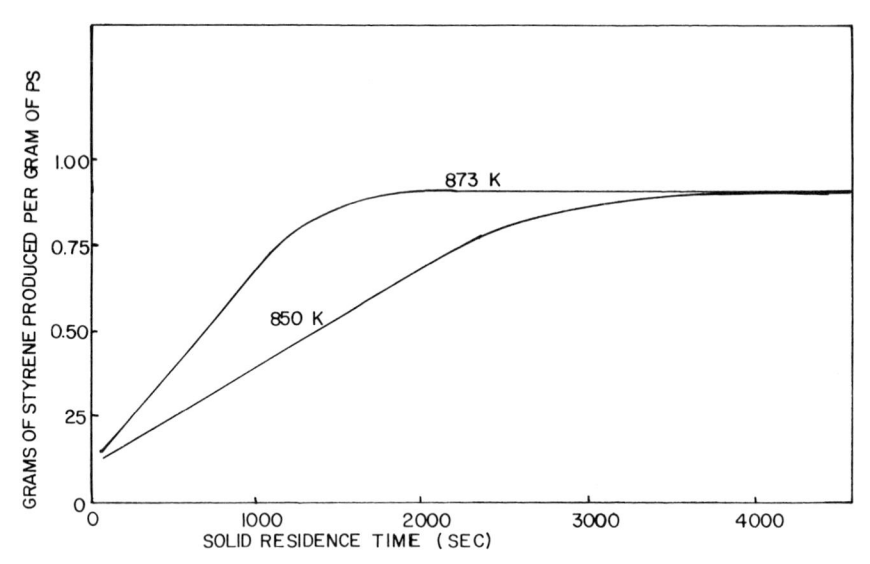

FIG. 6.13 Kinetics of styrene formation of 850 and 873 K. (Reprinted from Ref. 12.)

Tertiary Recycling   241

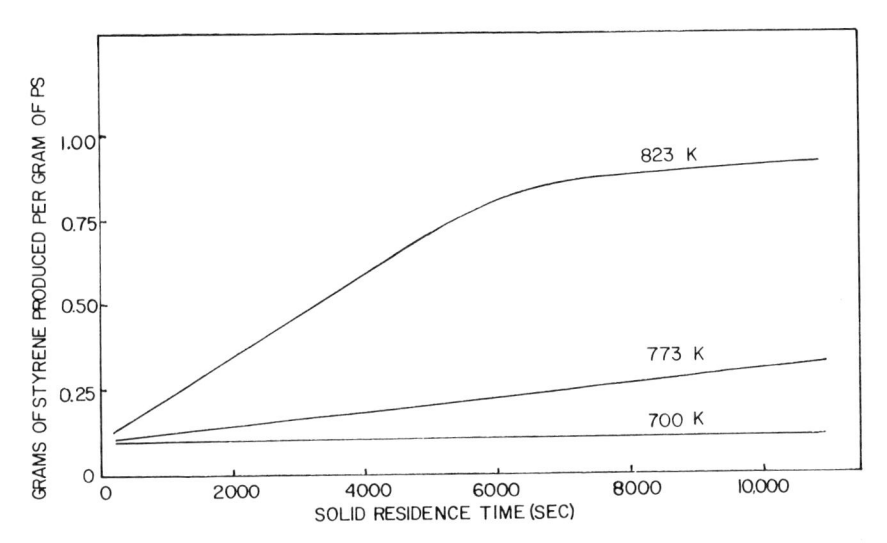

FIG. 6.14 Kinetics of styrene formation of 700, 773, and 823 K. (Reprinted from Ref. 12.)

The decomposition reaction mechanism of polyethylene proceeds in a different fashion than that of polystyrene. Reaction products resulting from the degradation of PE contain mainly paraffins (MW up to 50) rather than monomer. The degradation reactions at 660 to 710 K seem to be of zero order over a large range of weight loss. Once the chain is initiated, the mechanism probably consists of breaking off large molecular fragments in rapid succession. Figure 6.16 shows the kinetics of product formation furing the thermal decomposition of PE, and Fig. 6.17 gives feasible reaction path species. As with PS, PE equilibrium results indicate the formation of benzene at 773 K and 1 atm. The reaction intermediates are, however, different. In addition to toluene (present also in the decomposition of PS) the feasible intermediates include ethane, propane, ethyl bezene, and ortho-meta-para-dimethyl-benzene.

The thermal degradation of PVC is mainly by way of the dehydrochlorination reaction. After a brief initial period HCl

POLYSTYRENE

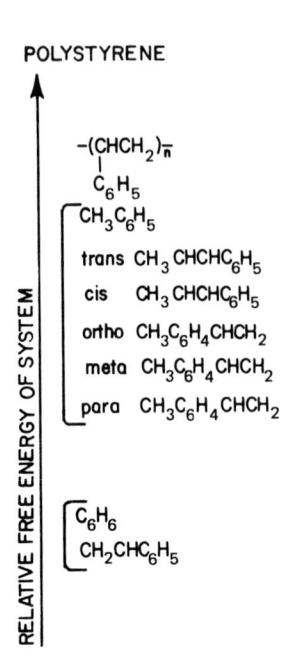

RELATIVE FREE ENERGY OF SYSTEM

$-(CHCH_2)_{\overline{n}}$
$|$
C_6H_5

$CH_3C_6H_5$
trans $CH_3CHCHC_6H_5$
cis $CH_3CHCHC_6H_5$
ortho $CH_3C_6H_4CHCH_2$
meta $CH_3C_6H_4CHCH_2$
para $CH_3C_6H_4CHCH_2$

C_6H_6
$CH_2CHC_6H_5$

FIG. 6.15 Thermal decomposition of PS: feasible reaction path species. (Reprinted from Ref. 12.)

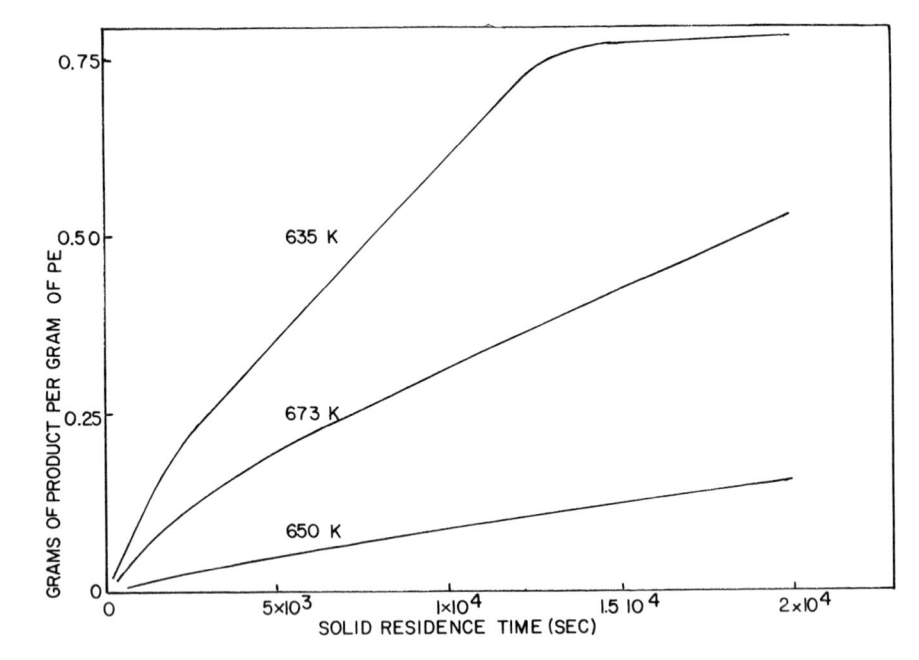

FIG. 6.16 Pyrolysis of PE at 635, 673, and 650 K: amount of product formed as a function of residence time. (Reprinted from Ref. 12.)

242

Tertiary Recycling 243

POLYETHYLENE

RELATIVE FREE ENERGY OF SYSTEM ↑

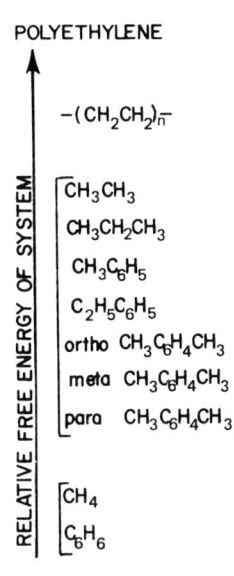

$-(CH_2CH_2)_{\overline{n}}-$

$\begin{bmatrix} CH_3CH_3 \\ CH_3CH_2CH_3 \\ CH_3C_6H_5 \\ C_2H_5C_6H_5 \\ \text{ortho } CH_3C_6H_4CH_3 \\ \text{meta } CH_3C_6H_4CH_3 \\ \text{para } CH_3C_6H_4CH_3 \end{bmatrix}$

$\begin{bmatrix} CH_4 \\ C_6H_6 \end{bmatrix}$

FIG. 6.17 Thermal decomposition of PE: feasible reaction path species. (Reprinted from Ref. 12.)

is evolved under a mechanism of approximately 3/2 order (Figs. 6.18 and 6.19). After HCl has been removed from PVC, a secondary decomposition takes place which yields a variety of organic products (Figs. 6.20 and 6.21). The feasible reaction path species are shown in Fig. 6.22.

The major equilibrium products of thermal decomposition of PVC at 773 K and 1 atm pressure are benzene and hydrogen chloride [11,12].

Figure 6.23 shows thermal decomposition characteristics of various plastics. Although most plastics decompose at the temperatures between 200 and 350°C, in order to achieve high decomposition rates, typical plastics pyrolysis reactors are designed to operate at much higher temperatures.

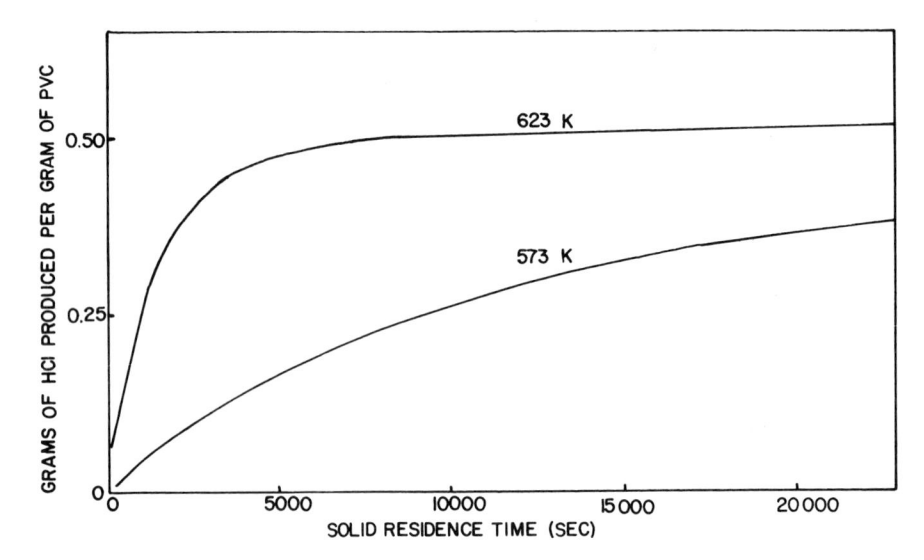

FIG. 6.18 Pyrolysis of PVC at 623 and 573 K: generation of HCl. (Reprinted from Ref. 12.)

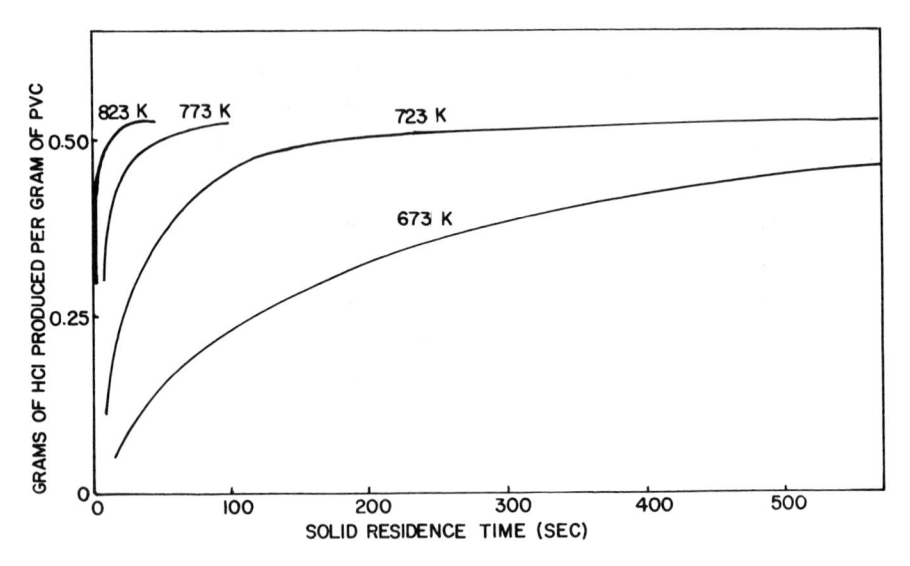

FIG. 6.19 Pyrolysis of PVC at 673, 723, 773, and 823 K: generation of HCl. (Reprinted from Ref. 12.)

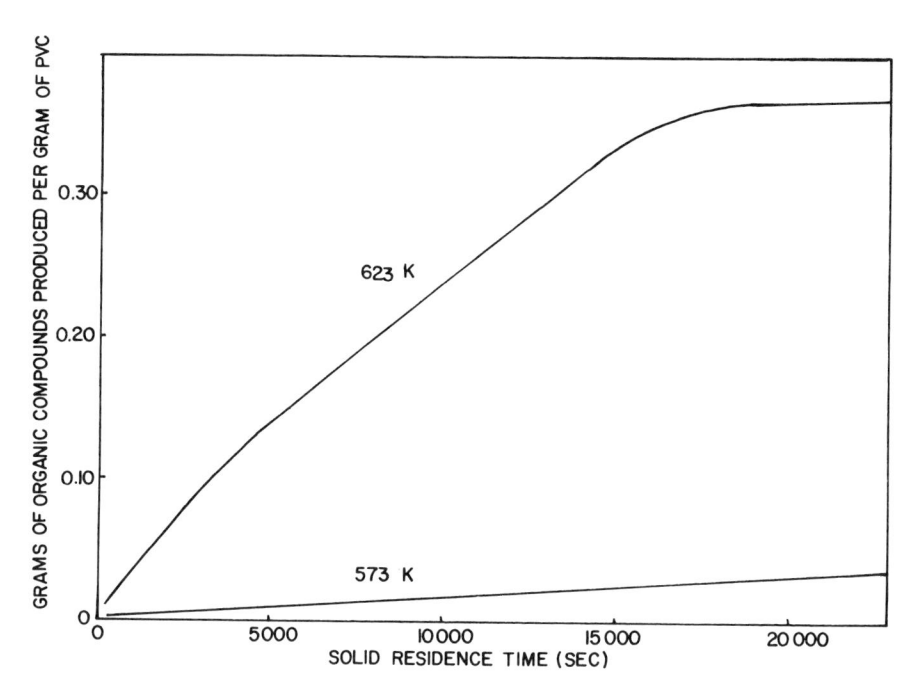

FIG. 6.20 Pyrolysis of PVC at 573 and 623 K: generation of organic compounds. (Reprinted from Ref. 12.)

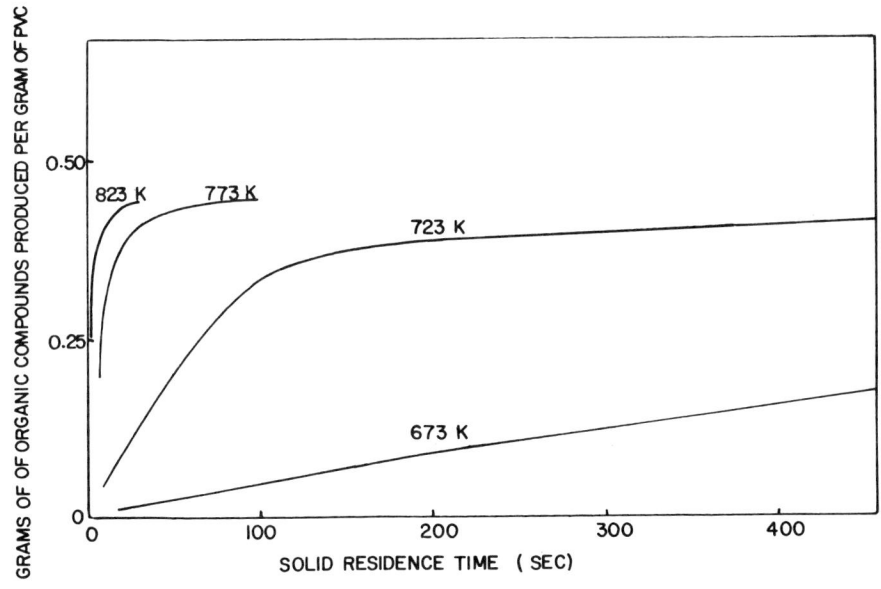

FIG. 6.21 Pyrolysis of PVC at 673, 723, 773, and 823 K: generation of organic compounds. (Reprinted from Ref. 12.)

POLYVINYL CHLORIDE

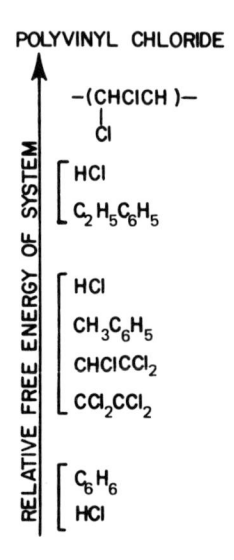

RELATIVE FREE ENERGY OF SYSTEM →

$-(CHClCH)-$
  $|$
  Cl

$\begin{bmatrix} HCl \\ C_2H_5C_6H_5 \end{bmatrix}$

$\begin{bmatrix} HCl \\ CH_3C_6H_5 \\ CHClCCl_2 \\ CCl_2CCl_2 \end{bmatrix}$

$\begin{bmatrix} C_6H_6 \\ HCl \end{bmatrix}$

FIG. 6.22 Thermal decomposition of PVC: feasible reaction path species. (Reprinted from Ref. 12.)

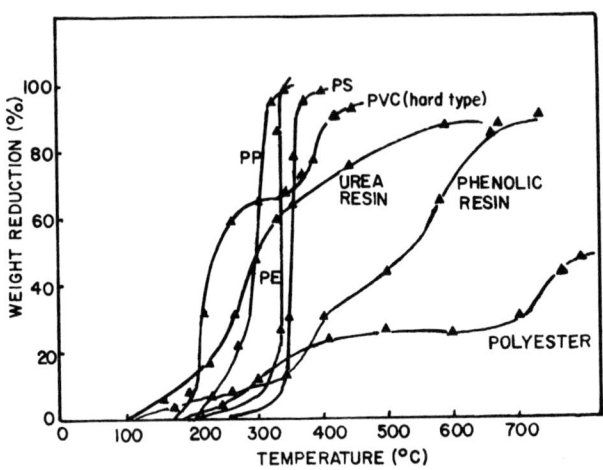

FIG. 6.23 Thermal decomposition characteristics of various plastics. (From Ref. 13.)

Tertiary Recycling   247

2. *Reactors for Pyrolysis of Plastics*

Because waste consisting mainly of plastics has different
handling properties from municipal refuse, pyrolysis of plastics
requires specialized equipment. The following problems are
encountered in the pyrolysis of plastics:

Since plastics have a poor thermal conductivity, a long time is
   required before thermal decomposition is reached, resulting
   in a small processing capacity as compared with the dimensions
   of the pyrolysis apparatus.
Carbon residue has a tendency to adhere to the walls of the
   pyrolysis reactor and is usually difficult to discharge
   continuously.
Many plastics on heating produce high viscosity melts which
   are difficult to transport.

Table 6.9 lists some of the plastics pyrolysis systems. Each
of these systems attempts to solve the problems just listed.

   a. *The Union Carbide System*

Union Carbide developed and tested a pyrolysis system consisting of an extruder, pyrolysis tube, heat exchanger, and
product recovery equipment (Fig. 6.24). A 1-1/4-in. electrically heated extruder is used to compress, melt, and pump

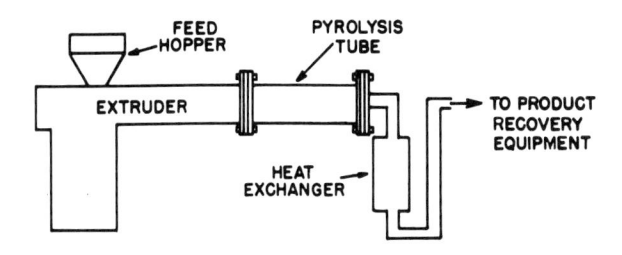

FIG. 6.24 Union Carbide's apparatus for the continuous pyrolysis
of plastics. (Reprinted from Ref. 14, courtesy American Chemical
Society.)

TABLE 6.9 Some Plastics Pyrolysis Systems

Process developer	Reactor type and heating method	Reaction temperature (°C)	Plant capacity (tons/day)	Feedstock	Products
Union Carbide	Extruder, followed by annular pyrol. tube, electrically heated	420-600	0.035-0.07	PE, PP, PS, PVC, PETP, PA, mixes	Waxes
Japan Steel Works	Extruder				
Japan Gasoline Co.	Tubular reactor, externally heated			Dissolved or suspended in recycle-oil	Heavy oil
Prof. Tsutsumi	Tubular reactor, superheated steam as a heat carrier	500-650	1	Ps-foam	
Sanyo Electric Co.	Tubular reactor with a screw for carbon removal, dielectric heating reactor	260 (PVC) followed by 500-550	0.3 (pilot) (Gifu) 3 (Kusatsu) 5	Foam PS, mixed plast. (select. collect.)	Monomer Fuel oil HCl
Mitsui Shipbuilding & Engineering Co.	Stirred tank reactor, polymer bath	420-455	24-30	Low-MW polymers (PE, APP)	Fuel oil
Mitsui Petrochemical Industries Co. (Chiba Works)					
Mitsubishi Heavy Ind. (Mihara Works)	Tank reactor with circulation pump and reflux cooling	400-500	0.7/2.4	Polyolefins	Naphtha kerosene fuel
University of Hamburg	Fluidized bed	640-840	0.08-0.24	PE, PS, PVC type rubber	Hydrocarbons
	Molten salt bath	600-800	Laboratory scale		Hydrocarbons

TABLE 6.10 Conditions for Pyrolysis of HDPE

Temperature (°C)				Throughput rate (g/hr)	Melt viscosity (140°C centipoise)
Zone 1	Zone 2	Pyrolysis tube	Heat exchanger		
350	460	437	215	2,700	304,000
350	460	457	205	2,580	32,000
320	475	465	150	2,500	3,000
350	460	505	195	1,800	450

Reprinted from Ref. 14, courtesy American Chemical Society.

molten polymer into the pyrolysis tube. The pyrolysis tube is of annular design in order to achieve relatively uniform temperature of the pyrolyzed plastic. Products are cooled in a heat exchanger before being discharged into the product recovery apparatus. Tables 6.10, 6.11, and 6.12 show operating conditions used in the pyrolysis of high-density polyethylene, low-density polyethylene, and polypropylene, respectively. Pyrolysis products of high-density polyethylene and polypropylene are hard waxes, superior in hardness and color to carnauba wax.

TABLE 6.11 Conditions for Pyrolysis of LDPE

Pyrolysis temperature (°C)	Throughput rate (g/hr)	% Yield[a]	Ring + ball[b] (m.p. °C)
555	2058	97	111.5
600	3216	97.5	107
600	2670	95.5	104
600	2124	91.0	60
625	2496	90.3	68
625	1944	88.5	58.5

[a] Wt % gross product not distillable at 160°C and 20 mm.
[b] Applies to product after distillation

Reprinted from Ref. 14, courtesy American Chemical Society.

TABLE 6.12 Conditions for Pyrolysis of PP

Temperature (°)				Throughput rate (g/hr)	Melt viscosity (170°C centipoise)
Zone 1	Zone 2	Pyrolysis tube	Heat exchange		
350	400	450	195	1500	4200
350	400	450	215	1800	1700
350	400	500	215	1860	520
350	400	525	230	1800	210

Reprinted from Ref. 14, courtesy American Chemical Society.

These have potential application in polishes, printing inks, release agents, and lubricants. Decomposition of low-density polyethylene results in the production of greases and waxes which can be made emulsifiable in water and used in polish formulations, textile finishing agents, adhesives, and lubricants. During the pyrolysis of polymers some degree of unsaturation is produced, which is useful in providing reactive sites for further chemical modification of the product. The apparatus developed by Union Carbide has also been used to pyrolize a mixture of various plastics. Depending on the temperature, tar, grease, liquids, or gases can be produced. Low-molecular-weight pyrolyzate is similar to petroleum (Table 6.13) and can serve as a petrochemical raw material source [14,15].

b. *The Japan Steel Works Ltd. System*

A pyrolysis system developed by Japan Steel Works Ltd., similar to Union Carbide's system, also utilizes an extruder. The pyrolysis, however, takes place directly in the extruder barrel rather than in a separate chamber (Fig. 6.25). Plastics are fed into the hopper (1), picked up by the rotating screw (2), plasticized, and thermally decomposed by the action of heat supplied both externally and generated by the shearing action of the screw. Low molecular weight

TABLE 6.13 Comparison of Petroleum Crude and Waste Plastic Pyrolyzate Produced by Union Carbide's Process

	Crude petroluem	Plastic pyrolyzate
Saturated hydrocarbon	Yes	Yes
Straight-chain	Yes	Yes
Branched-chain	Yes	Yes
Cyclic	Yes	Possibly traces
Unsaturated hydrocarbons	Yes	Yes
Straight-chain	Traces	High
Branched-chain	No	Yes
Aromatic	Yes	Yes
Alkylbenzene	Yes	Styrene, toluene
Styrene telomers	No	Yes
Naphthalenes	Yes	No
Phenanthrenes	Yes	No
Sulfur compounds	Yes	No

Reprinted from Ref. 14, courtesy American Chemical Society.

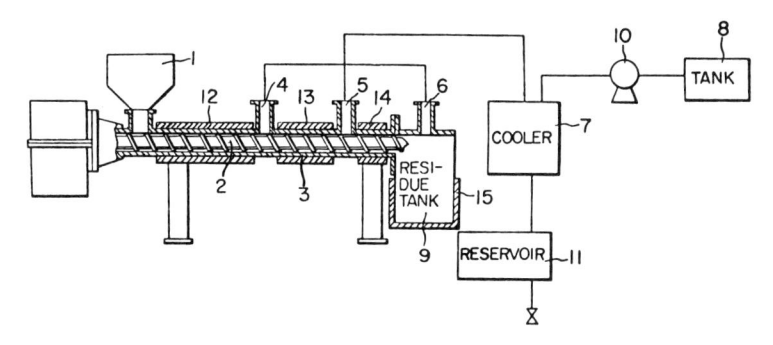

FIG. 6.25 Pyrolysis system developed by Japan Steel Works Ltd. (Reprinted from Ref. 16.)

TABLE 6.14 Operating Conditions and Yields of the Plastics Pyrolysis System Developed by Japan Steel Works Ltd.

Polymer	Thermal decomposition temperature (°C)	Degree of vacuum (gauge pressure; mmHg)	Number of rotations of screw (rpm)	Yield ratio of liquid products (%)	Amount of monomer in the liquid (%)
PS	500	-50	30	96.7	74.5
	500	-250	30	—	80.9
	500	-500	30	96.0	79.0
	600	Atmosphere	40	97.0	72.8
	600	-50	40	100.0	77.3
	600	-250	40	99.5	81.6
	600	-500	40	98.0	79.4
HIPS	500	-50	30	91.1	76.2
	500	-500	30	95.7	74.1
	500	-650	30	95.7	79.5

Tertiary Recycling

ABS	600	Atmosphere	60	78.0	89.3
	600	-250	60	96.5	76.0
	600	-500	60	97.9	78.6
	500	-100	50	67.8	51
	600	-100	50	88.8	44
PMMA	500	-100	50	99.6	Not specified
	500	-250	50	97.0	
PP	600	Atmosphere	40	70.0	Not specified
	600	-100	40	68.5	
	600	-250	50	72.5	
PE	600	Atmosphere	20	66.0	Not specified
	600	Atmosphere	40	76.0	
	600	-100	40	80.0	

Reprinted from Ref. 16.

products are fed to a cooler (7) through discharge orifices (4),
(5), and (6), then condensed, and the condensate is collected in a
reservoir (11). Some products having high boiling points can be
returned to the extruder for further reprocessing. Adhesion of
residue carbon to the inner surface of the cylinder can be prevented by the action of the rotating screw (3). Small amounts of
air also can be introduced into the decomposition zone to oxidize
the carbon residue and thus reduce the adhesion problem as well as
the amount of external heat necessary for the process. The mean
molecular weight of the product can be controlled by adjustment of
the decomposition temperature and by regulating the degree of
vacuum in the thermal decomposition zone by means of the vacuum
pump (10). Table 6.14 lists the operating conditions and yields
obtained in the pyrolysis of polystyrene, high-impact polystyrene,
acrylonitrile butadiene styrene, polymethylmethacrylate, polypropylene, and polyethylene [16].

   c. *The Mitsui Plastic Waste Thermal Cracking Process*
   Figure 6.26 shows a schematic of a thermal cracking process
developed by Mitsui Petrochemical Industries Limited and the Mitsui
Shipbuilding Engineering Company Limited. The feed for the process
is a low-molecular-weight polyethylene by-product from the HDPE
plant and atactic PP waste. The feed materials are melted and
pumped into a stirred tank reactor maintained at approximately
420°F. The products of pyrolysis are distilled off: part of the
entrained mist and some high-molecular-weight products are condensed and returned to the reactor. Product oil is mixed with
heavy fuel and used in the power plant; the carbon residue is
continuously removed. The required amount of heat of decomposition is obtained by intermittently drawing off the fluid from the
decomposition process and combusting it in a burner in the combustion chamber. The process fluid drawn from the reactor
amounts to about 10% of the feedstock used [17,18].

Tertiary Recycling 255

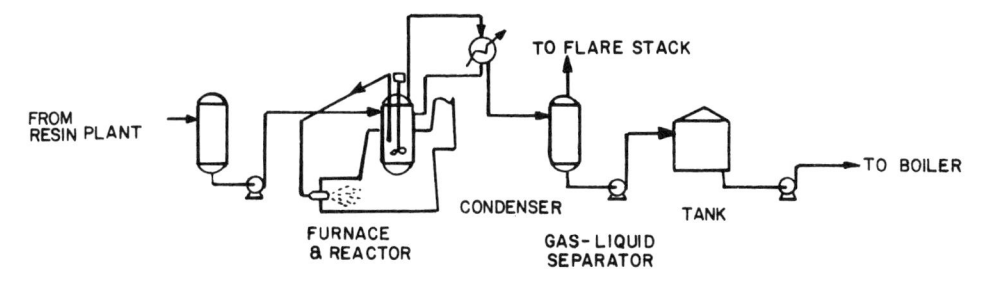

FIG. 6.26 Mitsui's thermal cracking process. (From Ref. 17.)

d. *Mitsubishi's Thermal Decomposition Apparatus*

The thermal decomposition process developed by Mitsubishi Heavy Industries Limited is shown schematically in Fig. 6.27. The material is crushed to uniform size, dried, charged into a feed hopper, and fed into the melting vessel. The charge is melted at 300°C, at which temperature HCl is removed from the PVC and goes to a condenser. The molten plastic is then fed into the decomposition vessel. The pyrolysis reactor has provisions for the internal reflux of the products of pyrolysis (Fig. 6.28).

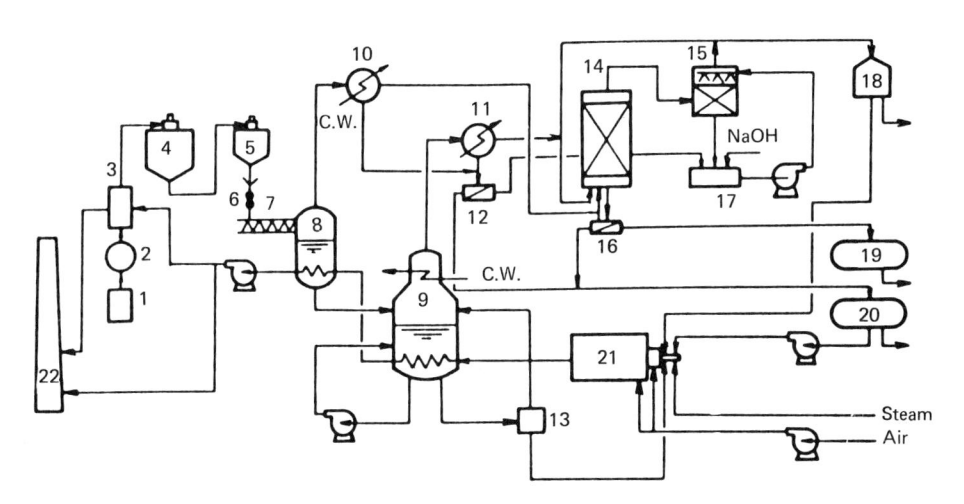

FIG. 6.27 Schematic of Mitsubishi Heavy Industries Ltd.'s plastics waste pyrolysis system. (From Ref. 13.)

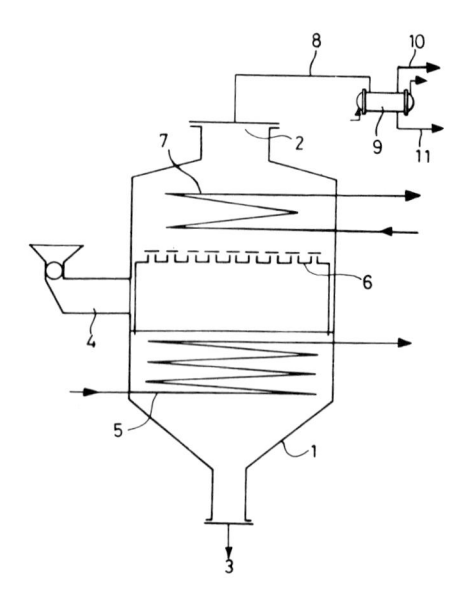

FIG. 6.28 Pyrolysis reactor developed by Mitsubishi Heavy Industries Ltd. (Reprinted from Ref. 19.)

The plastic waste is charged into the pyrolysis chamber (1) from a feeder hopper (4). The decomposition products are subjected to gas-liquid contact in the tray-shaped pack (6) (similar to the packs used in the distillation columns), and then pass through the cooling region, which is kept at a desired temperature by means of the gas cooling tube (7). Heavy products are condensed and fall on the tray-pack where they contact the ascending gases, and are then required to the decomposition zone. The products produced in this apparatus are characterized by narrow boiling point distribution. For heating the decomposition vessel, high-temperature gas is generated by burning part of the product. The exhaust gas from the decomposition vessel is utilized in the melting and drying process. Table 6.15 shows typical products obtained in this process [13,19].

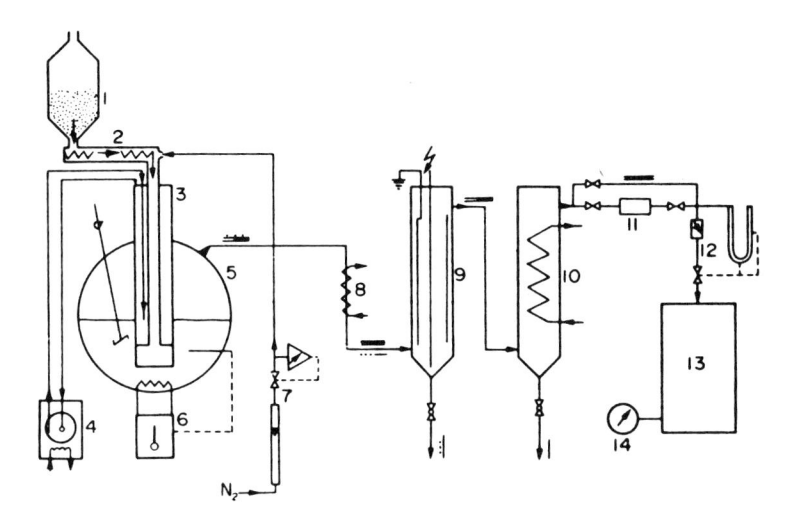

FIG. 6.29 University of Hamburg molten salt plastics pyrolysis apparatus. (Reprinted from Ref. 21, courtesy Prof. Dr. W. Kaminsky.)

e. *Hamburg University Molten Salt Reactor*

Figure 6.29 is a schematic of an apparatus for the pyrolysis of plastic waste in molten salt. The plastics are transported into the heated melt by a screw. After decomposition, the hot vapors pass through an electrostatic precipitator where paraffin vapors are condensed to form relatively pure paraffin waxes. The liquid fraction is separated from the hydrocarbon gas in the intensive cooler. Table 6.16 gives material balances in the pyrolysis of PE, PVC, and a mixture of PE, PVC, and PS. In the pyrolysis of PE, the yield of ethylene and methane increases and the yield of propene decreases with increasing temperature. At 850°C ethylene and propene are the main products. Hydrogen, ethane, propane, isobutane, and butadiene are present only in small amounts. The amount of aromatic compounds increases with increasing temperatures, but so does the formation of char. If polystyrene is pyrolyzed at temperatures between 650 and 700°C, a large amount of styrene is produced. When the temperature is increased above

TABLE 6.15 Products of Mitsubishi Thermal Decomposition Process

Kinds of materials processed		Product composition (wt %)			Yield (%)	Flash pt. (°C)	Viscosity (cst)
		Gas	Oil	HCl			
PP	(1)	1.8	97.1		98.9	-15	1.21
PP	(2)	4.3	95.0		99.3	-15	1.10
PP PS PE	50% 20% 30%	5.4	93.8		99.2	-30	1.16
PP Phenol	75% 25%	5.3	82.9		88.2	-6	1.46
PE PVC	75% 25%	1.1	71.6	10	82.7	3	1.42
LDPE HDPE PP PS Others	35.7% 14.2% 10.8% 16.9 4.4%	11.9	73.2	7.7	92.8	2	1.33

Reprinted from Ref. 13.

700°C, there is a sharp decrease in benzene, methane, and ethylene, and an increase in char formation. Pyrolysis of PVC yields large amounts of HCl and a mixture of hydrocarbons [20,21].

f. Hamburg University Fluidized Bed Reactor

The layout of the fluidized bed reactor (Fig. 6.30) used in the plastic pyrolysis studies at Hamburg University is similar to that of the molten salt reactor. The screw transfers plastic from the storage vessel to the electrically heated reactor. The fluidized layer is about 80 mm deep, and requires 500 liters of fluidizing

Characteristics of product oil

Pour pt. (°C)	Ash (%)	Sulfur (%)	Spec. wt.	Cal. val. (kcal/kg)	Moisture (%)
Less than -40	Less than 0.01	Less than 0.01	0.766	10.960	Less than 0.1
Less than -40	Less than 0.01	Less than 0.01	0.766	11,170	Less than 0.1
Less than -40	Less than 0.01	Less than 0.01	0.802	10,620	Less than 0.1
Less than -40	Less than 0.01	Less than 0.01	0.778	11,270	Less than 0.1
-4	0.03	Less than 0.01	0.782	11,680	Less than 0.1
-4	Less than 0.01	Less than 0.01	0.796	11.210	Less than 0.1

gas per hour. The pyrolysis gases are cleaned of dust in the cyclone. The paraffin mist is separated by the electric filter. The gases are then partially liquified in intensive coolers. The uncondensed gases can be returned for use as fluidizing medium. Table 6.17 gives the results of the pyrolysis of PE, PS, PVC, and a plastics mixture. In all cases it was possible to recover over 97% of the original plastics feedstock. The main product of the pyrolysis of PE is ethylene. The amount of benzene depends on whether nitrogen or cracked gas is used as fluidized medium. In the pyrolysis of PS, styrene monomer is the main product. The pyrolysis of PVC yields up to 50% wt/wt of hydrogen

TABLE 6.16 Materials Balance Using Molten Salts as Heating Medium

Input material:	PE				PVC				PS			
Temperature (°C):	640	690	740	790	850	640	690	740	790	60%/40% 790	640	690	740	490

Yield (wt %)													
Hydrocarbon gases to C_4 and H_2	40.6	56.1	62.0	65.9	63.2	4.7	5.0	5.6	6.6	37.8	1.8	2.2	5.7	9.5
HCl	—	—	—	—	—	57.9	58.1	55.4	57.2	22.7	—	—	—	—
Carbon	0.05	0.2	1.2	2.3	7.9	7.3	8.9	9.8	12.1	9.8	3.2	3.4	7.2	17.9
Liquid and solid products except carbon	56.2	42.6	33.7	31.3	27.0	29.0	26.3	25.2	24.0	28.1	93.5	92.9	83.8	72.4
Materials balance	96.9	98.9	96.9	99.5	98.1	98.9	98.3	96.0	99.9	98.4	98.5	98.5	96.7	99.8
Portion not measurable by GLC except carbon[a]	26.1	12.1	10.2	3.6	3.5	14.0	14.0	12.1	8.9	9.7	0.05	2.0	6.3	13.8

[a]The molecular weights of some of the compounds formed were so high that they could not be measured by GLC.
Reprinted from Ref. 21, courtesy Prof. Dr. W. Kaminsky.

Tertiary Recycling

TABLE 6.17 Composition of Pyrolysis Products in the Laboratory Fluidized Bed Reactor (% w/w)

Product:	PE	PE	PE	PS	PVC	Mixture
Fluidizing medium:	N_2	Cracking gas	C-gas, zeolite	Cracking gas	Cracking gas	Cracking gas
Temperature (°C)	1013	1013	1063	1013	1013	1063
Hydrogen	0.3	0.5	1.9	0.03	0.7	0.7
Methane	7.0	16.2	16.7	0.3	2.8	13.2
Ethylene	35.1	25.5	10.3	0.5	2.1	13.2
Ethane	3.6	5.4	4.1	0.04	0.4	2.0
Propene	22.6	9.4	6.4	0.02	0	0.1
Isobutene	8.7	1.1	2.3	0	0	0.1
1,3-Butadiene	10.3	2.8	2.5	0	0	0.7
Pentene and hexene	0.01	2.0	6.1	0.01	0	0.6
Benzene	0.01	12.2	7.4	2.1	3.5	14.7
Toluene	0.05	3.6	51.1	4.5	1.1	45.
Xylene and ethyl benzene	0	1.1	3.3	1.2	0.2	0.9
Styrene	0	1.1	0.6	71.6	0	10.5
Naphthalene	0	0.3	0.8	0.8	3.1	2.5
Higher aliphates and aromatics	10.53	17.3	12.1	15.2	19.3	19.2
Carbon	0.4	0.9	18.3	0.3	8.8	2.9
Chlorinated hydrocarbons	0	0	0	0	56.3	8.1
Hydrogen sulfide	0	0	0	0	0	0
Fillers	0	0	0	0	0	0
Total	98.6	99.4	97.9	96.6	98.7	97.3

Reprinted from Ref. 22, courtesy Kunststoffe.

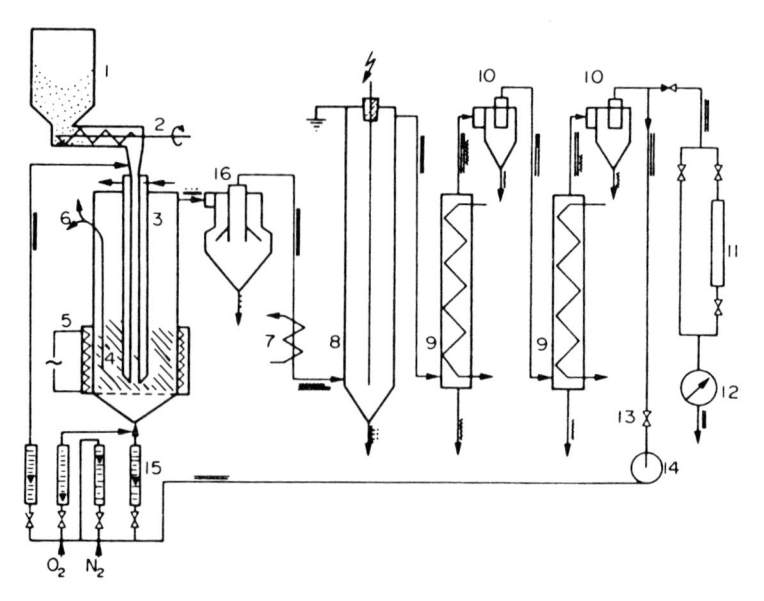

FIG. 6.30 University of Hamburg experimental fluidized bed pyrolysis reactor. (Reprinted with permission from Ref. 20, copyright 1976, Pergamon Press, Ltd.).

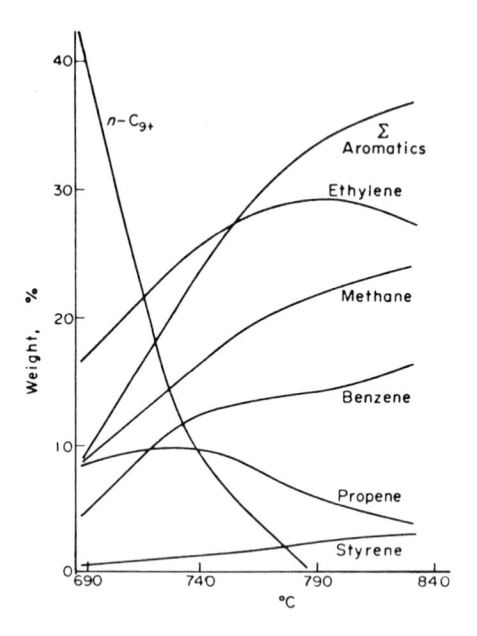

FIG. 6.31 Pyrolysis of PE in fluidized bed reactor with recycle gas as fluidizing medium: amounts of products as functions of temperature (Reprinted with permission from Ref. 20, copyright 1976, Pergamon Press, Ltd.)

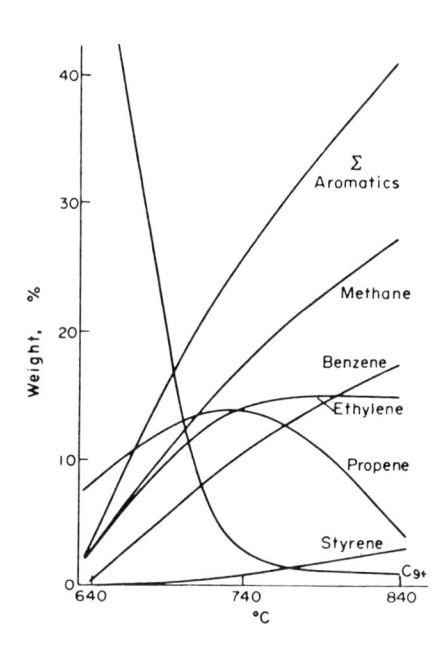

FIG. 6.32 Pyrolysis of PP in a fluidized bed reactor with recycle gas as fluidizing medium: amounts of products as functions of temperature. (Reprinted with permission from Ref. 20, copyright 1976, Pergamon Press, Ltd.)

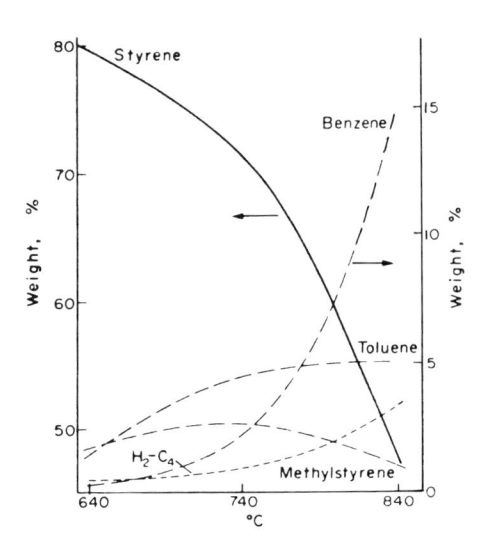

FIG. 6.33 Pyrolysis of PS in fluidized bed reactor with recycle gas as fluidizing medium: amounts of products as function of temperature. (Reprinted with permission from Ref. 20. Copyright 1976, Pergamon Press, Ltd.)

chloride gas and considerable amounts of carbon [20-22]. The effects of temperature on pyrolysis yields are shown in Figs. 6.31, 6.32, and 6.33.

g. Other Systems

The Sanyo Electric Company developed a pyrolysis plant in which the rapid heating of plastics to the melting point is achieved by a microwave heater [20,23]. The Japan Gasoline Corporation developed a fluidized bed recycling process in which polystyrene wastes are fed into a fluidized bed which is stirred and heated by the preheated fluidization medium and by partial oxidation. The yield of reprocessible styrene monomer is 50 to 60% of the feed [20]. Tsutsumi [23] proposed a pyrolysis system using superheated steam as a cracking medium. In this process the pyrolysis reactor would be operated near the solid waste incinerator: the steam produced by the incinerator would be additionally heated to approximately 600°C and used in the pyrolysis of plastics. Guzzeta et al. [24] described a process for depolymerizing polystyrene with a molecular weight of over 1 million, using heat and substances which yield stable free radicals. These free radicals combine with polystyrene free radicals resulting from scission of the chains, to form inactive compounds which do not join together to produce branches or cross-links. Albert and Tacke [25] developed a process to recover polyesters from scrap polyurethanes by pyrolysis. The reclaiming process is carried out by igniting scrap foam in a vessel having a perforated bottom and side walls. As the resin foam burns, a liquid consisting essentially of polyesters drains through the perforations and is collected. Practically all the polyester portion is recovered, and can be used in combination with new ingredients for the production of polyurethane foam.

II. CHEMICAL DECOMPOSITION OF PLASTICS WASTE

Decomposition of waste plastics by chemical rather than thermal means is possible for a variety of plastics. The chemical decomposition of plastics has certain advantages over pyrolysis: composition of the product is more uniform and easier to control; usually less separation

and purification of products is required; lower capital investment is
needed; and the recovery plant can operate economically at considerably
smaller throughput. The main drawback of chemical process is that
they require relatively clean and uniform feed and are unsuitable for
mixed plastics waste. Although it is possible to decompose many
polymers, the main interest today is in polyurethanes and to a smaller
extent, thermoplastic polyesters. These are of particular interest
because of the large quantities available: polyurethane foam scrap
generated during manufacture of the products and from car shredding
operations, and thermoplastic polyester scrap from bottles, X-ray
films, and textiles.

A. Hydrolysis

Breaking down by hydrolysis is possible for plastics containing
chemical groups susceptible to such reactions. Since hydrolysis is
a reverse reaction to condensation, it is those polymers obtained
by polycondensation (or polyaddition) that can be hydrolyzed. Examples of such polymers are polyurethanes, polyesters, polycarbonates, and polyamides. Since plastics are made to be hydrolytically
stable under normal usage conditions, the hydrolysis of plastics
waste has to be carried out under extreme conditions. Of the hydro-
lyzable plastics, the recycling of polyurethanes is of greatest
industrial interest.

The reaction of polyurethanes with superheated steam was studied
by Leverkusen [26] and Mahoney et al. [27]. Polyurethane hydrolyzes
according to the general equation,

$$-NH-\overset{\overset{O}{\|}}{C}-X + H_2O \rightarrow -NH_2 + CO_2 + H-X \tag{3}$$

where X represents either a polyol or -NH-linkage. Upon mixing low-
density polyurethanes with superheated steam for about 15 min the
foam is converted to a liquid denser than water. In the specific case
described [27] the liquid contains 65 to 85% theoretical yield of
toluene diamines and 90% yield of polypropylene oxide. The formation of toluene diamine was shown to be pseudo-first-order in the

temperature range 160 to 190°C:

$$+\frac{d(TDA)}{dt} = K' - NH-\overset{\overset{O}{\|}}{C} \qquad (4)$$

where $NH-\overset{\overset{O}{\|}}{C}-$ represents the total amount unreacted urea and polyetherurethane linkages [27]. The production of toluene diamenes at various temperatures is shown in Fig. 6.34 at 160°C. There is a slight deviation from linearity at low conversions, caused by isomeric diamine being formed as a result of a variety of different groups hydrolyzing at different rates.

Out of the two hydrolysis products, the polyol portion (polyester or polyether) can be directly used in a new foam formulation, while the amine portion has to be chemically converted to isocyanate. Table 6.18 compares the data on virgin polyether used to manufacture

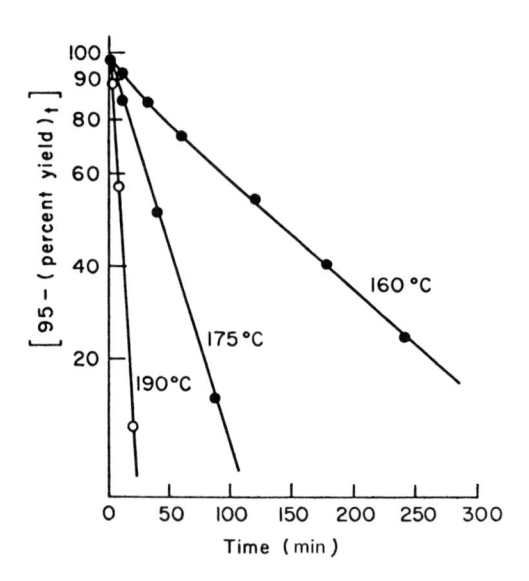

FIG. 6.34 Hydrolysis of flexible polyurethane foam: kinetics of the formation of toluene diamines. (Reprinted with permission from L. R. Mahoney, S. A. Weiner, and F. C. Ferris, "Hydrolysis of Polyurethane Foam Waste," Environmental Science and Technology 8(2), 1974.

TABLE 6.18 Comparison of Virgin and Recovered Polyether

	Specification desmophen 7100	Recovered polyether average values
Hydroxyl value	49 ± 3	50.9
Acid value	<0.01	0.63
Water content (%)	<0.1	0.065
Viscosity (cP at 25°C)	520 ± 30	1.4551
Ash (%)	<0.05	0.05
Unsaturated compounds (mVal/g)	<0.05	0.095
Peroxide content (ppm)	<30	5
pH value	6.6-7.2	6.0
Pour point (°C)	<-40	<-50

Reprinted from Ref. 26, courtesy Kunststoffe.

the polyurethane foam and polyether recovered from foam by hydrolysis. Both products are very similar, and when recycled polyether is used in typical formulations a usable foam is produced (Tables 6.19 and 6.20).

Figure 6.35 schematically shows a flowsheet of the polyurethane foam hydrolysis plant designed by General Motors. Scrap foam enters the reactor where it hydrolyzes in contact with steam at approximately

TABLE 6.19 Foam Formulations Using Virgin and Recycled Polyether

Test	Standard polyether	Recycled polyether	H_2O	Catalyst A	B	C	TDI
1	100	—	4.0	0.5			
2	90	10	4.0	0.5			
3	80	20	4.0	0.5			
4	50	50	4.0	1.0			
5	—	100	4.0	1.2			
6	—	100	4.0	1.2			

Reprinted from Ref. 26, courtesy Kunststoffe.

TABLE 6.20 Foaming Characteristics and Foam Properties of Formulations Using Virgin and Recycled Polyether

Test	Cream time (sec)	Fill rise time (sec)	Fiber time (sec)	Air permeability (mm water column)	Density (kg/m³ DIN 53420)	Tensile strength (kPa)	Elongation at break (DIN 53571) (%)	Stiffness in compression at 40% compression (kPa) (DIN 53577)
1	9	86	28	150	27	110	190	4.1
2	6	82	41	160	25/26	105	170	4.0/3.9
3	7	113	84	210	26/26	95	155	3.9/3.9
4	7	83	91	220	27/27	85	135	3.8/3.9
5	6	54	61	250	23/23	95	180	3.7/3.7
6	5	57	54	230	25/25	105	190	4.0/4.1

Reprinted from Ref. 26, courtesy Kunststoffe.

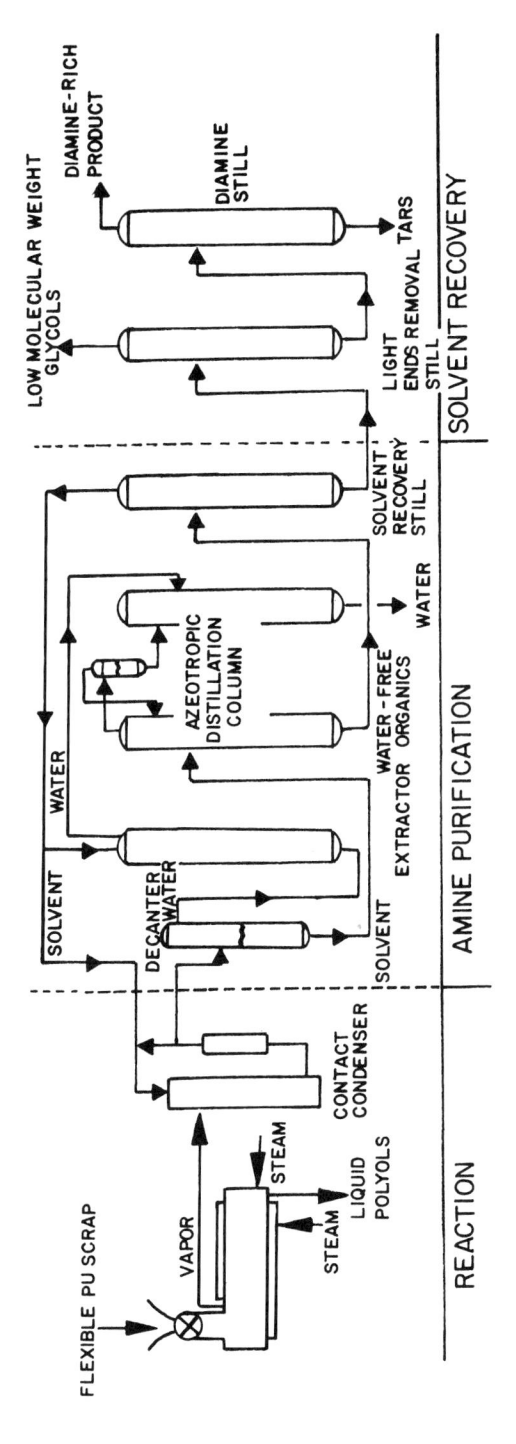

FIG. 6.35 Flow sheet of General Motors' polyurethane hydrolysis plant. (Reprinted from Ref. 28, courtesy McGraw-Hill.

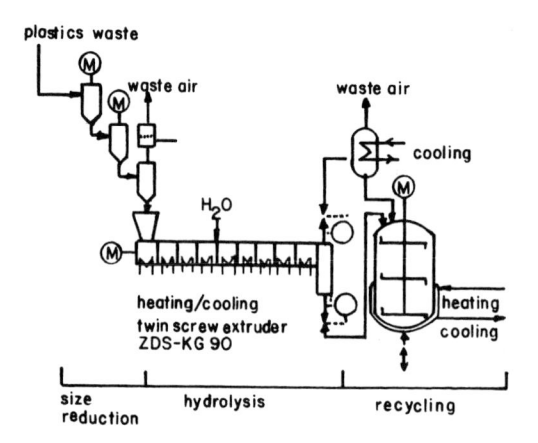

FIG. 6.36 Continuous hydrolysis apparatus. (Reprinted from Ref. 26, courtesy Kunststoffe.)

600°F. Polyols are recovered directly as liquids which are relatively water-free and ready for reuse after cooling and filtering. Vapor from the reactor passes into a spray condenser where it contacts an aniline or benzyl alcohol solvent. Various solvent-recovery processes separate water, solvent, and organic products. Distillation separates the organic stream into diamines, the main product, and glycol and tar by-products [28].

Leverkusen [26] designed a continuous hydrolysis reactor, utilizing a twin-screw extruder as a reaction chamber (Fig. 6.36). The extruder is built to withstand temperatures of up to 300°C and the resulting pressure, and to give a residence time of 5 to 30 min. A feed screw compresses the shredded foam and passes it into the hydrolysis zone. The water required for hydrolysis is pumped countercurrently to the direction of the foam. Hydrolysis converts the compressed solid into a pulp, which is thoroughly wetted by the water with the help of the kneading discs on the screw. Hot, hydrolyzed material leaves the extruder through a controlled pressure relief valve. The product consists mainly of polyether and amine from the isocyanate. Separation of the two products can be accomplished in one of three ways: (1) removal of amine by distillation, followed by purification of polyether; (2) precipitation of the amine (by reaction with acid) followed by filtration of the precipitate; and, (3) addition of a second liquid phase that would dissolve only one component [26].

B. Glycolysis

The necessity of separating amines and glycols produced by the hydrolysis of polyurethanes is a serious drawback to the hydrolysis process. This drawback can be eliminated by a polyurethane recovery process based on glycolysis. Glycolysis is a relatively simple process in which polyurethane foam is decomposed at 185 to 200°C in the presence of an appropriate glycol. The chemistry of the glycolysis process involves transesterification of the carbonate groups in the polyurethane foam with glycol solvent. The general reaction is shown in Fig. 6.37. As can be seen, the reaction results in only one product, a polyol mixture. No separation is necessary, and the recovered mixed polyols can be reused in combination with virgin material [29,30].

Table 6.21 compares the properties of a typical rigid foam formulation based on virgin glycol with mixtures of virgin and recycled glycols. Glycol recovered by glycolysis produces foams practically indistinguishable from those made with virgin glycol. Glycolysis can be used for decomposing rigid and flexible polyurethane as well as isocyanurate foams.

The industrial glycolysis process is very simple. Precut or pulverized foam is fed into a heated reactor containing glycol at 185 to 210°C under a nitrogen atmosphere. The rate at which the foam is fed into the reactor is a function of the type of agitation and mass, and the heat transfer. A good mixing is necessary since the foam does not wet easily and has a tendency to float on

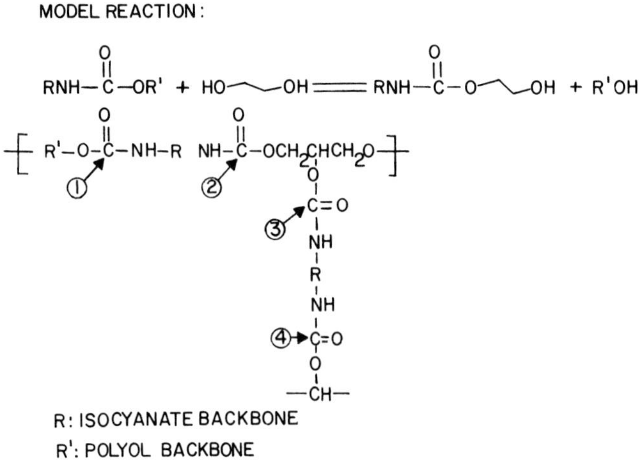

MODEL REACTION:

$$RNH-\overset{O}{\overset{\|}{C}}-OR' + HO\frown OH \rightleftharpoons RNH-\overset{O}{\overset{\|}{C}}-O\frown OH + R'OH$$

$$\left[R'-O-\overset{O}{\overset{\|}{C}}-NH-R \quad NH-\overset{O}{\overset{\|}{C}}-OCH_2CHCH_2O \right]$$
①   ②   O
                        ③ C=O
                           NH
                           R
                           NH
                  ④ C=O
                         O
                       -CH-

R: ISOCYANATE BACKBONE
R': POLYOL BACKBONE

FIG. 6.37 Chemistry of glycolysis of polyurethane foam. (Reprinted from *Advances in Urethane Science and Technology*. Technomic Publishing Co., Inc., 265 Post Road West, Westport, Conn. 06880.)

the surface of the glycol. The cost of producing recycled polyol is low enough to make the process economically attractive. Glycolysis of polyurethane wastes may be catalyzed by certain organometallic compounds and/or tertiary amine catalysts.

C. Other Processes

Broeck and Peabody [31] describe a process for reclaiming polyurethane foam by dissolving it in liquid prepolymer and using the solution to produce a new foam. The dissolving of the foam is carried out at temperatures approximating 250°C. It is possible to produce mixtures containing up to 50% dissolved scrap, but concentrations below 25% are recommended.

Heiss [32] developed a process for the liquefaction of scrap polyurethanes and polyureas by dissolving them in carboxylic acids having molecular weights below 5000 and acid numbers from 50 to 800. The rate with which the scrap dissolves depends on the relative proportions of carboxylic acid, and urethane or urea

TABLE 6.21 Rigid Foams Based on Virgin and Recovered Polyols

	Parts by Weight for Different Foams						
	A	B	C	D	E	F	G
Components							
PAPI	140	140	140	140	140	140	140
Aromatic tetrol	100	60	60	60	60	60	60
Recycled polyol from foams based on the following polyols:							
Aromatic tetrol		40					
Sorbitol			40				
Pentaerithritol				40			
Methylglycoside					40		
Sucrose						40	
Aromatic pentol							40
Properties							
Density (pcf)	2.03	2.09	2.09	2.02	2.00	2.00	1.96
Compression (%)	17.3	21.5	20.9	18.8	19.8	17.6	16.1
Oxygen index (%)	23.0	24.5	24.3	24.0	24.3	24.5	24.5

Reprinted from *Advances in Urethane Science and Technology*, Technomic Publishing Co., Inc., 265 Post Road West, Westport, Conn. 06880.

groups. The higher the acid number in proportion to the number of urethane or urea groups present, the faster the reaction at a given temperature. Examples of reagents which may be used in the process are fatty acids, resin acids, polymerized unsaturated acids, polycarboxylic acids, heterocyclic acids, and the mixture containing suitable acids such as tall oil, stearine pitch, fish oil pitch, and linseed pitch. The products obtained can be used in coatings, adhesives, and foams.

Matsudaira et al. [33] described a process for the recovery of polyethers and diamino-toluenes from waste polyurethane foam or elastomers by thermal decomposition at temperatures below 250°C in the presence of an aliphatic amine, a cycloaliphatic amine, an aromatic amine, a heterocyclic amine, or derivatives of these amines, and one of the following types of compounds: an alkali metal oxide, an alkali metal hydroxide, an alkaline earth metal oxide, an alkaline earth metal hydroxide, or an aqueous solution of these compounds.

The decomposition of polyurethane results in the formation of two liquid immiscible layers: an upper layer containing crude polyether and a lower one containing diamine. Recovered polyether can be reused in the manufacture of foams, while the diamine can be converted to isocyanate and then used in foam formulation.

MacDowell [34] describes a process for the reclamation of waste terephthalate polyester films or fibers, by heating the waste with glycol and thus converting it to diglycol terephthalate. This monomer can then be added to the virgin monomer and used to polymerize new polyester resin.

Gruschke et al. [35] developed a two-stage process for depolymerization of polyethylene terephthalate. The first stage is carried out in an autoclave in which polyester waste is melted and mixed with 2 to 8 times the amount of methanol. The residence time in the autoclave is limited to approximately 10 min. The second stage is carried out under the same pressure conditions and at slightly lower temperatures in a reaction tube without stirring. After the reaction is completed, the solution of diethyl terephthalate in methanol is transferred to a stirred vessel having an internal temperature of 100°C, and the pressure is reduced to atmospheric. In most cases the decomposition product is so pure that polyethylene terephthalate or degradation products of low molecular weight cannot be detected by the usual analytical methods.

Stevenson [36] describes a process for the recovery of high purity dicarboxylic acid and glycol from scrap polyesters containing dyes and other impurities. The process is based on the reaction of waste polyester in the form of fibers, films, pellets, molded pieces, and so on with an alcohol in an autoclave at an elevated temperature, followed by catalytic hydrogenation. Depolymerization products, while in solution, may be purified by filtering or by the absorption of impurities on activated charcoal. After depolymerization and purification, the material is subjected to a selective hydrogenation conducted under such conditions that unsaturated color-forming impurities are hydrogenated while the main product is not. The unsaturated color-forming impurities thus become colorless. The hydrogenation process, in the presence of metal catalysts such as palladium, platinum, and ruthenium, is carried out at temperatures between 50 and 200°C and pressures ranging from atmospheric to 2500 psig.

REFERENCES

1. *Pyrolysis*, Fact Sheet from the National Center for Resource Recovery, Inc., Washington, D.C., March 1973.
2. P. R. Bell and J. J. Varjavandi, "Pyrolysis-Resource Recovery from Solid Waste," *Australian Waste Management & Control Conference,* University of South Wales, Kensington, 1974.
3. T. Lamb, "Pyrolysis Fundamental Methodology," *1975 Int. Symp. on Energy Recovery from Solid Waste*, Kentucky Center for Energy Research, Lexington.
4. W. S. Sanner, C. Ortuglio, J. G. Walters, and D. E. Wolfson, *Conversion of Municipal and Industrial Refuse into Useful Materials by Pyrolysis*, Report of Investigation 7428, U.S. Bureau of Mines, Washington, 1970.
5. C. J. Huang and C. Dalton, *Energy Recovery from Solid Waste*, NASA-CR-2526, 1975.
6. J. L. Pavoni, J. E. Heer, Jr., and D. J. Hagerty, *Handbook of Solid Waste Disposal, Materials and Energy Recovery,* Van Nostrand Reinhold, New York, 1975.
7. T. Szebely, J. J. Fritz, and F. A. Berczynski, "The Andco-Torrax Slagging Pyrolysis Solid Waste Disposal System," *Third National Chemical Engineering Conference* at Mildura, Victoria, Australia, Aug. 1975.
8. D. A. Tillman, "Fuels from Recycling Systems, *Env. Sci. Technol. 9(5),* 1975.
9. "Disposing of Solid Wastes by Pyrolysis," *Env. Sci. Technol. 9(2),* 1975.

10. G. A. Zerlaut and A. M. Stake, "Chemical Aspects of Plastic Waste Management," *Recycling and Disposal of Solid Waste,* Teh. F. Yen, (Ed.), Ann Arbor Science Publishers, Ann Arbor, 1974.

11. R. S. Ottinger, M. E. Banks, and W. D. Lusk, "Utilization of Waste Plastics Through New Chemical Concepts," Presented before the Division of Water, Air and Waste Chemistry, American Chemical Society, Chicago, Sept. 1970.

12. M. E. Banks, W. D. Lusk, and R. S. Ottinger, *New Chemical Concepts for Utilization of Waste Plastics,* Report SW-1GC for the U.S. Environmental Protection Agency, 1971.

13. K. Matsumoto, S. Kurisu, and T. Oyamoto, "Development of Process of Fuel Recovery by Thermal Decomposition of Waste Plastics," *Int. Conf. on Conversion of Refuse to Energy,* Montreaux, Switzerland, 1975.

14. J. E. Potts, "Reclamation of Plastics Waste by Pyrolysis," Presented before the Division of Water, Air and Waste Chemistry, American Chemical Society, Chicago, 1970.

15. J. E. Potts, "Continuous Pyrolysis of Plastic Wastes," *Ind. Water Eng. 7(8),* 1970.

16. H. Tokushige, A. Kosaki, and T. Sakai, "A Method for the Continuous Thermal Decomposition of Synthetic Macro-molecular Materials," U.K. Pat. 1,369,964.

17. Y. Kitaoka and H. Sueyoshi, "Conversion of Waste Polymer to Fuel Oil," *Int. Conf. on Conversions of Refuse to Energy,* Montreaux, Switzerland, 1975.

18. H. Sueyoshi and Y. Kitaoka, "Make Fuel from Plastic Waste," *Hydrocarb. Proc.* Oct. 1972.

19. S. Kurisu, T. Oyamoto, S. Ochiai, H. Miyamoto, and K. Makino, "Thermal Degradation of Organic Wasters," U.K. Pat. 1,413,870.

20. W. Kaminsky, J. Menzel, and H. Sinn, "Recycling of Plastics," *Cons. Rec. 1,* 1976.

21. J. Menzel, H. Perkow, and H. Sinn, "Recycling Plastics," *Chem. Ind.* June 16, 1973.

22. W. Kaminsky and H. Sinn, "Pyrolysis of Plastics Waste and Used Tyres in a Fluidized Bed Reactor," *Kunststoffe 68(5),* 1978.

23. S. Tsutsumi, "Thermal and Steam Cracking of Waste Plastics," *Int. Conf. on Conversion of Refuse to Energy,* Montreaux, Switzerland, 1975.

24. G. Guzzetta, F. Sabbioni, and G. B. Gechele, "Polymerization of Styrene in the Presence of Free Radical Generating Substance," U.S. Pat. 3,143,536, 1964.

25. H. Albert and W. Tacke, "Method for Reclaiming Scrap Polyurethane Resin," U.S. Pat. 2,998,395, 1964.
26. E. G. Leverkusen, "Hydrolysis of Plastics Wastes," *Kunststoffe 68(5)*, 1978.
27. L. R. Mahoney, S. A. Weiner, and F. C. Ferris, "Hydrolysis of Polyurethane Foam Waste," *Env. Sci. Technol. 8(2)*, 1974.
28. "New Routes Tackle Tough Plastics-Recycling Jobs," *Chem. Eng.*, Feb. 16, 1976.
29. H. Ulrich, "Recycling of Polyurethane and Isocyanurate Foam," *Adv. Ureth. Sci. Technol. 5*, 1978.
30. H. Ulrich, A. Odinak, B. Tucker, and A. A. Sayigh, "Recycling of Polyurethane and Polyisocyanurate Foam," *Polym. Eng. Sci., 18(11)*, 1978.
31. T. R. Broeck and D. W. Peabody, "Method for Reclaiming Cured Cellular Polyurethanes," U.S. Pat. 2,937,151, 1960.
32. H. L. Heiss, "Method of Dissolving Polyurethanes and Polyureas Using Tall Oil," U.S. Pat. 3,109,824, 1964.
33. N. Matsudaira, S. Muto, Y. Kubota, T. Yashimoto, and S. Sato, "Method of Decomposing Urethane Polymer," U.S. Pat. 3,404,103, 1966.
34. J. T. MacDowell, "Process of Reclaiming Linear Terephthalate Polyester," U.S. Pat. 3,222,299, 1965.
35. H. Gruschke, W. Hammerschick, and H. Medem, "Process for Depolymerizing Polyethylene Terephthalate to Terephthalic Acid Dimethylester," U.S. Pat. 3,403,115, 1968.
36. G. M. Stevenson, "High Purity Polyester Depolymerization Products from Polyester Scrap by Polish Hydrogenation," U.S. Pat. 3,501,420, 1970.

Chapter 7

QUATERNARY RECYCLING: ENERGY FROM PLASTICS WASTE

I. INTRODUCTION

Incineration of refuse may be defined as the reduction of combustible wastes to inert residue by controlled high-temperature combustion. The main reason for incineration is reduction in the volume of waste. Incineration is capable of reducing the weight of refuse by 80% and the volume by over 90%. The residue from the refuse is inert and may be disposed of in landfill. Because of recent rapid increases of the cost and energy, more attention has been focused on the possibility of utilizing energy from the combustion of solid municipal and industrial refuse. Since paper constitutes a considerable portion of municipal solid waste, its heating value is relatively high. A typical unprocessed municipal refuse has a calorific value of 4000 to 6000 Btu/lb as compared to 11,000 to 12,000 Btu/lb for coal. The calorific value of the refuse can be increased by the removal of some noncombustible components. Energy recovery from municipal solid refuse can take the following routes:

Burning Refuse in Steam-generating Incinerators. The heat generated during incineration produces steam which may be used to heat and air-condition buildings, or for industrial processing or the production of electricity.

Burning Refuse in Existing Heat Exchangers. Refuse can be used to supplement fossil fuels in existing power boilers.

Pyrolysis. By pyrolyzing refuse, a transportable fuel is produced (see Chap. 6).

Hydrogenation. Refuse can be converted to heavy oil by heating it under pressure in the presence of carbon monoxide and steam.

Anaerobic Digestion. In this process the organic portion of the refuse is decomposed in the absence of oxygen. Methane produced has the potential of being used as a natural gas substitute [1].

At the present time, burning refuse in steam-generating incinerators and its use as supplemental fuel are the most advanced waste energy utilization technologies. The use of refuse as a supplemental fuel in power-producing plants offers the following advantages: (1) maximum heat is recovered from the refuse; (2) air pollution emissions are lower than in most other waste incineration processes; (3) nonrenewable natural resources are conserved; (4) power production costs are reduced; and (5) the process is an environmentally acceptable method of solid waste disposal [2]. The advantages of energy recovery in specially designed incinerators are (1) well-designed incinerator can consume various kinds of refuse; (2) a reasonably centrally located site is possible, reducing transportation costs; and (3) incinerators are flexible in handling varying amounts of refuse. The disadvantages of incineration are (1) a large capital investment is required; (2) operating cost is relatively high; (3) there are difficulties in obtaining sites for incineration plants, especially in the proximity of residential areas; and (4) air emissions must be controlled [3].

II. THE INCINERATOR

A municipal refuse incinerator receives municipal refuse and reduces its volume by up to 90% by combustion with the possible recovery of heat.

A typical incinerator comprises a set of scales, a storage pit and tipping area, incinerator cranes, charging mechanisms, a furnace, and pollution control devices.

A. Scales

Scales are used to determine the weight of an incoming feed. Weight records can be used to improve incinerator operation, to assist in management control of the facility, to assist in planning for new facilities, and to provide means for assessing fees for processing. In small incinerators the scales may be manually operated, but in large units they must be automated [2].

B. Storage Pit and Tipping Area

The storage pit into which trucks deposit the incoming refuse has to be large enough to hold an adequate amount of refuse. Continuous incinerators require smaller storage pits than the batch type. Both storage pit and tipping area have to be enclosed against wind and weather. Dumping doors or gates should be used to confine odors, dust, and noise produced during the unloading of vehicles. Storage pits are usually constructed of concrete, are 20 to 40 ft wide by 40 to 80 ft deep, and may be several hundred feet long. Water supply fixtures are provided to facilitate cleaning of the bottom and side walls, and to extinguish fires caused by spontaneous combustion [2].

C. Incinerator Cranes

Since in most incinerators the storage pit is deep, an overhead crane is required to transfer the refuse to the entrance of the charging facility. The number of cranes and the type used depend on the capacity of the incinerator units.

D. Charging Mechanisms

Most small incinerators are charged in batches. The waste is picked up from the storage pit by the crane and deposited in the charging gate or hatch. The gate is closed except when the waste is being charged. In the continuous feeding system solid waste is fed directly into the combustion chamber, usually through a vertical or inclined rectangular chute which is maintained full to create an air seal. Continuous feeding offers the advantages of being able to minimize irregularities in the combustion process and minimize the thermal shock caused by an addition of cold feed; and it is much easier to control than a batch operation.

Most incinerators are provided with charging hoppers built in the shape of an inverted truncated pyramid made of reinforced concrete or steel. Because of the need for good abrasion resistance, steel is the preferred material of construction [2].

E. Furnaces

Furnaces are generally constructed of concrete foundations with either a structural steel framework supporting the inner walls and roof arch of refractory material, or typical masonry with bricks laid one atop another and with self-supporting arched roofs. In the design of the furnace walls the emphasis is on withstanding high temperatures (in excess of 2000°F) and on preventing degradation (softening, erosion, slagging) of the refractory during operation. The furnace can be water-walled. The heat recovery is then accomplished by the use of furnace walls made of closely spaced steel tubes welded together. Water or steam circulates through the tubes and extracts heat from the combustion zone. Both the water-walled furnace with integral boiler or the refractory-lined furnace with a boiler mounted in the outlet flue can be used for energy recovery. Following are the common types of incinerator furnaces:

Rectangular Furnace (Fig. 7.1). This type of furnace can be used for continuous or batch operations. The refuse moves from the drying area through the ignition zone and into the combustion chamber. Mechanical motion of the transfer gates and the passage of air through the grates agitates the wastes. Combustion gases pass into the secondary combustion chamber, where more complete oxidation takes place. Solid residue passes out of the furnace on a conveyor [2].

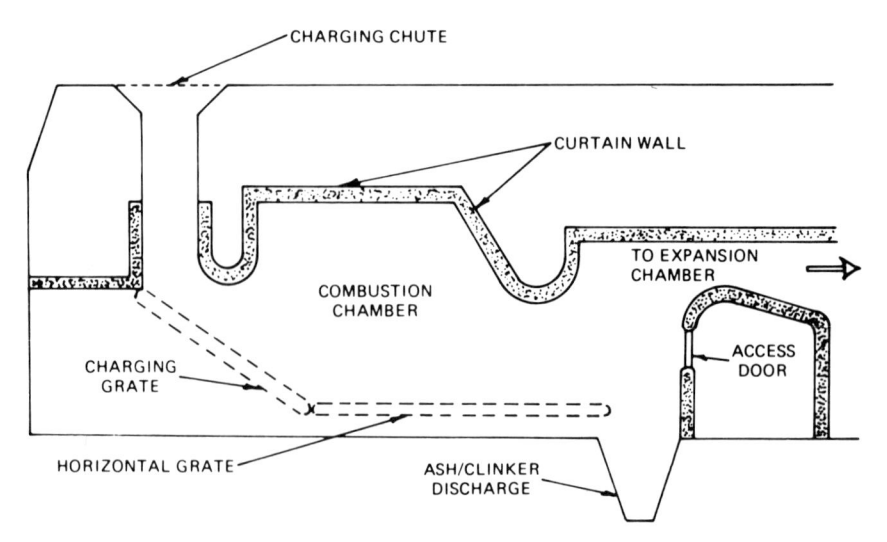

FIG. 7.1 Rectangular furnace. (From *Handbook of Solid Waste Disposal* by Joseph L. Pavoni, John E. Heer, and D. Joseph Haggerty, Copyright 1975 by Litton Educational Publishing, Inc. Reprinted by permission of Van Nostrand Reinhold Company).

Cylindrical Furnace (Fig. 7.2). The cylindrical furnace is often used in batch operations. Refuse is charged through a gate in the upper portion of the furnace, and, when the charging gate opens, is dropped onto the central cone-shaped grate. The cone-shaped grate is surrounded by a circular grate. The central grate rotates and arms attached to it move over the circular grate. The residue is gradually swept from the combustion process to the outside of the circular grate where it drops into a residue removal device. The primary combustion gases and particulate matter enter the secondary chamber where they are more completely oxidized [2].

Rotary Kiln Furnace. The rotary kiln furnace consists of a rectangular drying and ignition zone, and rotating combustion kiln.

Quaternary Recycling

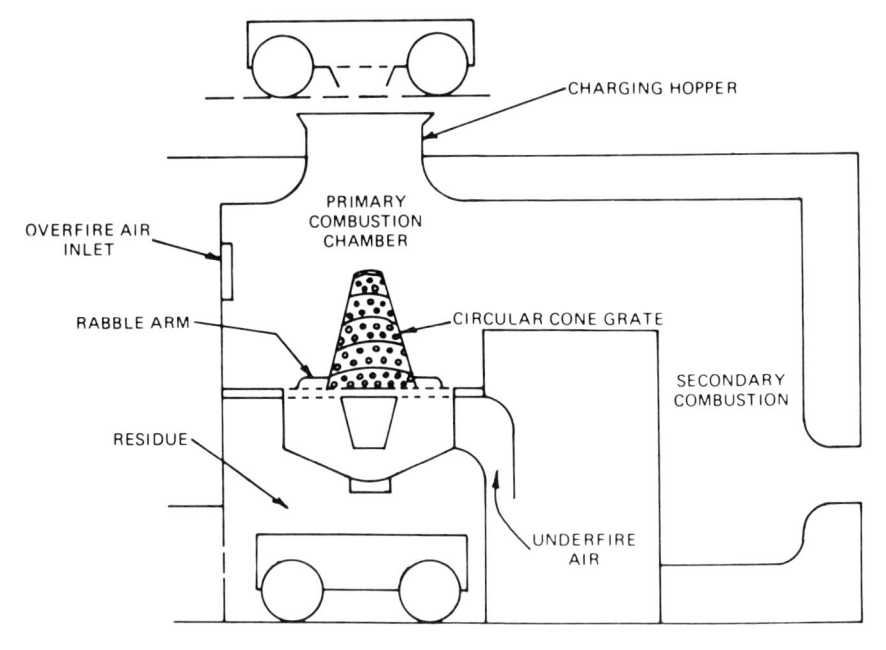

FIG. 7.2 Cylindrical furnace. (From Handbook of Solid Waste Disposal by Joseph L. Pavoni, John E. Heer, and D. Joseph Haggerty, Copyright 1975 by Litton Educational Publishing, Inc. Reprinted by permission of Van Nostrand Reinhold Company.)

The kiln is a large metal cylinder lined with refractory material with its axes slightly inclined. It is mounted on rollers and slowly rotated by electric motors. The refuse is first passed through the drying and ignition zones where moisture and volatile constituents are driven off. The final combustion takes place in the kiln. The volatiles driven off in the ignition zone go through a passage above the kiln and join the hot gas effluent from the kiln in a secondary combustion chamber where the combustion is completed. The solid residue moves downward along the bottom of the kiln and falls out of the end of the kiln into the refuse collecting device [4].

F. Pollution Control Devices

The flue gases contain solid and gaseous pollutants. The concentration of sulfur oxides is usually small compared to gases from oil or coal burning. Formation of hydrochloric acid may be a problem. HCl may be removed by scrubbing the gases in water. Particulates can be collected using mechanical (centrifugal) collectors, electrostatic precipitators, wet scrubbers, and filters.

III. EXAMPLES OF ENERGY RECOVERY FROM MUNICIPAL REFUSE

A. Supplemental Fuel

The Union Electric Co. was the first U.S. investor-owned utility to burn municipal solid refuse as supplementary fuel for the direct production of electric power.

The important part of the process is the preparation of feedstock (Fig. 7.3). Refuse is shredded and magnetic materials are removed. An air classifier further removes heavy and noncombustible materials. The light fraction, shredded to about 1.5-in. diameter, is transported from the preparation facility to Union Electric's plant (Fig. 7.4). The refuse is unloaded from the trucks to a receiving bin supplying the belt conveyor. A 12-in.-diameter pneumatic feeder pipe transports the refuse to a 8600-ft^3 storage site, a cyclone separating the air from the solids. Four conveyors feed the refuse to pneumatic feed systems for transport to the four refuse burners, where the refuse is burned in tangentially fired boilers in suspension with pulverized coal. The problems encountered during the initial stages of the operation were erosion of the piping of the conveying system, poor combustion of the fuel, increased bottom ash, and corrosion of the pneumatic refuse feeder system. In spite of these problems the experience gained by the Union Electric Co. indicates that almost any fossil fuel-fired boiler possessing ash-handling capability can be adapted for burning prepared refuse [2,3].

Quaternary Recycling 285

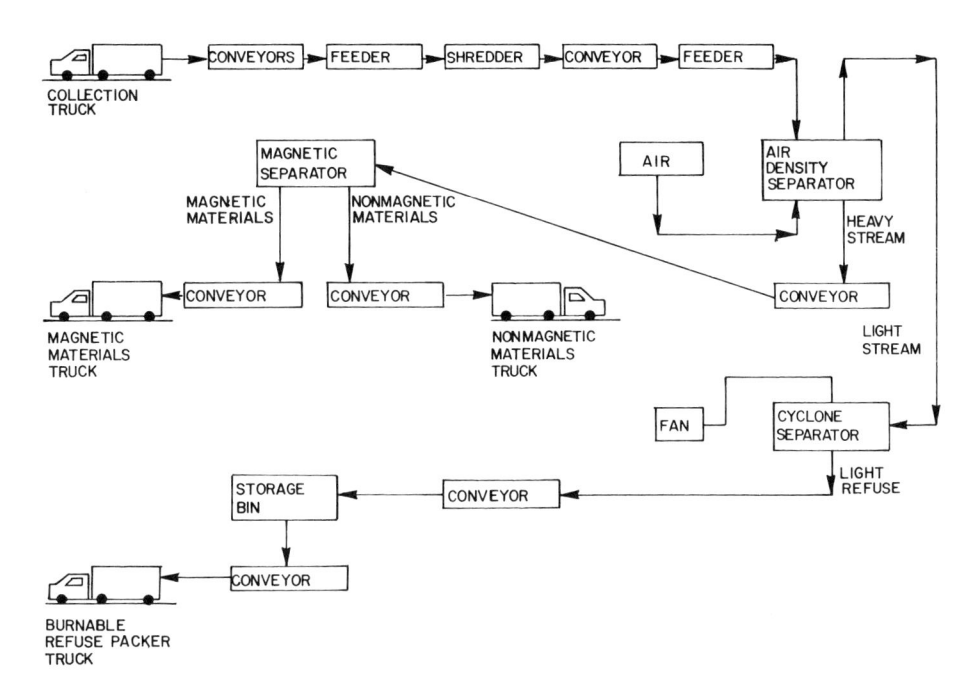

FIG. 7.3 Schematic of preconversion plant for Union Electric's supplemental fuel project. (Reprinted from Ref. 3, courtesy NASA.)

B. Incinerators with Energy Recovery Systems

The incinerator designed to burn 1008 metric tons per day of refuse collected from communities around Saugus, Massachusetts, is an example of a "refuse to energy" conversion plant, a schematic of which is given in Fig. 7.5. Trucks entering the plant are weighed and directed to the unloading areas. The crane deposits the refuse in an incinerator hopper, where it forms a seal, preventing flame flashbacks. From the hopper the refuse is fed to the incinerator grates, while combustion air is supplied to the underside of the grates. While moving down the first grate the refuse is dried and ignited. Most of the combustion takes place on the second grate, but the complete burnout occurs on the third grate. The solid residue falls through the grates, and is collected and discharged, while the flue gases,

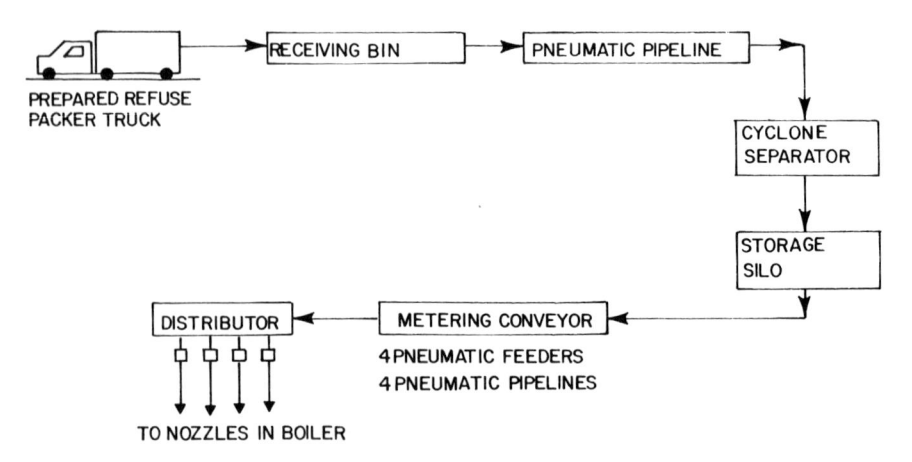

FIG. 7.4 Schematic of refuse-firing facilities at Union Electric's supplemental fuel plant. (Reprinted from Ref. 3, courtesy NASA.)

after leaving the furnace, pass through the boiler section and to precipitators. The steam produced is transferred to the General Electric power plant [3].

Combustion Power's CPU 400 incinerator system is another example of an incinerator especially designed to convert municipal refuse

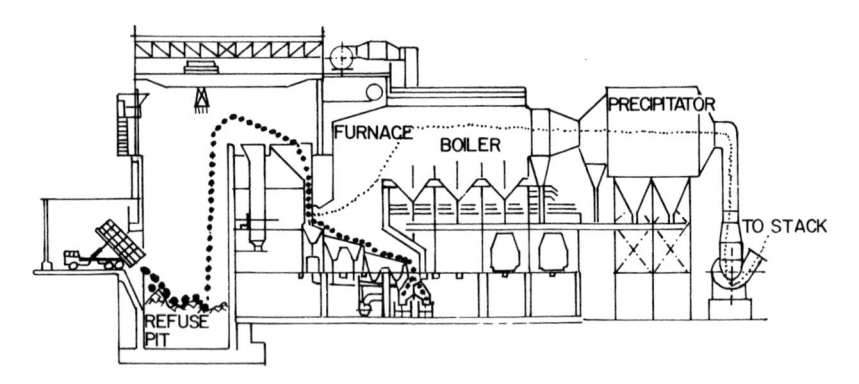

FIG. 7.5 Schematic of water wall incinerator at Saugus, Mass. (Reprinted from Ref. 3, courtesy NASA.)

Quaternary Recycling

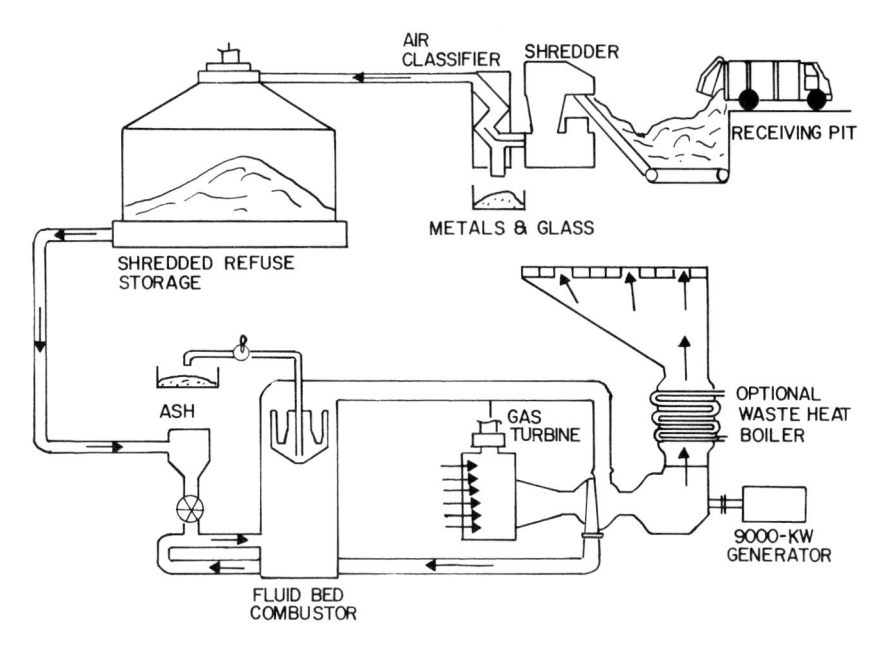

FIG. 7.6 Flowchart of the CPU-400 system. (Reprinted from Ref. 7, courtesy National Center for Resource Recovery.)

into energy. The schematic of the process is shown in Fig. 7.6. Instead of producing steam the system uses combustion gases to power a gas turbine coupled with an electric generator. Unseparated refuse is fed into the shredder and conveyed through an air classifier. The light fraction, mainly paper and plastics, is transported to the refuse storage bin. The light fraction from the storage container is continuously injected into the fluidized bed furnace, where it burns while being kept in suspension by air blowing from the underside. The pressurized gaseous products of combustion are cleaned and used to power the gas turbine generator, which can be put rapidly on and off line. A number of such installations can be integrated into an existing system with the fossil fuel plant carrying the base load [1-3].

IV. THE EFFECT OF PLASTICS IN MUNICIPAL REFUSE ON THE INCINERATION PROCESS

Kaiser and Carotti [5] investigated the effect on incineration of plastics mixed with municipal refuse. In that study the combustion of refuse was done in the incinerator at the town of Babylon, Long Island, N.Y. The refuse alone was incinerated as a reference, followed by the combustion of refuse with 2 and 4% plastic added to it. Four plastics were tested: polyethylene, polystyrene, polyurethane, and polyvinyl chloride. All the tests were performed under normal operating conditions.

The combustion performance during the testing of PE is shown in Table 7.1. The addition of PE did not cause any considerable change in the low smoke density. Burnout occurred at a greater distance from the end of the grate after the addition of PE, indicating more rapid and complete burnout. The presence of an increased amount of PE resulted in a lower odor of flue gas.

An analysis of noxious chemical species in the flue gas is shown in Table 7.2. To eliminate the effect of variable dilution with air, the results were corrected to standard concentration (12% CO_2). The addition of PE caused a decrease in the concentration of organic acids in the flue gas, probably because of the higher combustion temperature. The variation in chloride, fluoride, and phosphate ions, and in sulfur dioxide, reflects the normal variations in refuse analysis. The higher combustion temperature caused by the addition of PE resulted in an increased concentration of nitric oxide.

The addition of PS gave similar results (Table 7.3). Less odor and no smoke were detected. No melted plastic was found in the grate siftings and no clogging of grates by plastics was observed. The concentrations of noxious species in the flue gas are shown in Table 7.4. Since PS contains less sulfur than the rest of the refuse, the addition of PS caused a decrease in SO_2 concentration in the flue gas. The cause of the increased concentration of organic acids is not understood. The increased NO concentration is

TABLE 7.1 General Combustion Performance During the Polyethylene Series

	Test no.		
	A-1	A-2	A-3
PE added (% refuse)	None	2	4
Refuse (lb/hr)	15,660	14,667	14,333
PE	0	293	573
Total (lb/hr)	15,660	14,960	14,906
Furnace arch (°F)	1,689	1,687	1,709
Flue gas (°F)	1,315	1,160	1,200
Flue gas analysis (Orsat: dry vol %)			
CO_2	6.48	5.2	5.7
CO	0.0	0.0	0.0
O_2	13.81	15.2	14.2
N_2	79.71	79.6	80.1
Burnout from end of grate (ft)	8.13	8.93	12.1
Smoke (Ringelmann)	0.014	0.10	0.03
Odor units of flue gas	2.0	1.5	1.0

Reprinted from Ref. 5. Study sponsored by Society of Plastics Industry.

within normal experimental variations. Table 7.5 shows the general combustion results for the municipal refuse with 2 and 4% of PU added. The low arch and flue-gas temperatures are the results of an unevenly wet and dry condition of the base refuse. PU foam burned rapidly. Some light char was lofted by the air and continued to burn in suspension. The somewhat higher smoke density might have been caused by the addition of the foam or by the lower combustion temperature. No evidence of melted PU was found. The concentrations of noxious chemical species in the flue gases are shown in Table 7.6. The concentration of chlorides increased. The NO concentration slightly increased, indicating that some of the PU nitrogen might

TABLE 7.2 Concentrations of Noxious Chemical Species in the Flue Glass During the Polyethylene Series (ppm by Volume of Dry Gas, Actual and Corrected to 12% CO_2)

	Test no.		
Chemical	A-1 No PE added	A-2 2%	A-3 4%
Cl^-, Actual[a]	217	126	201
Corrected	402	291	424
F^-, Actual	2.6	3.5	5.7
Corrected	3.8	8.1	10.5
CN^-, Actual	<0.02	<0.01	<0.01
Corrected	<0.04	<0.02	<0.02
NO, Actual	28.6	40.1	39.9
Corrected	53.0	92.7	84.0
PO_4^{3-}, Actual	0.54	0.54	0.45
Corrected	1.00	1.25	0.95
SO_2, Actual	31.7	35.4	33.2
Corrected	58.7	81.6	70.3
Organic acids, as CH_3COOH, Actual	58.3	17.1	5.5
Corrected	108	39.4	11.6
Aldehydes, as HCHO			
Actual	1.5	1.7	1.2
Corrected	2.8	3.9	2.5

Note: Tests were not run for phosgene and free chlorine, as the polyethylene contained no chlorine.
[a]Average of scrubber and NaOH/impinger methods.
Reprinted from Ref. 5. Study sponsored by Society of Plastics Industry.

have undergone oxidation. Tests on the municipal refuse alone and with PVC added were run simultaneously in two furnaces in order to reduce the experimental variations caused by the nonuniformity of the base

TABLE 7.3 General Combustion Performance During Polystyrene Series

	Test no.		
	B-1	B-2	B-3
PS added (% refuse)	None	2	4
Refuse (lb/hr)	16,333	14,544	14,333
PS	0	291	573
Total (lb/hr)	16,333	14,835	14,906
Furnace arch (F)	1,665	1,701	1,717
Flue gas (°F)	1,181	1,075	1,159
Flue gas analysis (Orsat; dry vol %)			
CO_2	6.12	2.11	2.01
CO	0.0	0.0	0.0
O_2	13.93	13.98	14.19
N_2	79.95	79.91	79.80
Burnout from end of grater (ft)	14.7	14.6	13.3
Smoke (Ringelmann)	0.00	0.00	0.00
Odor units of flue gas	2.0	1.4	1.4

Reprinted from Ref. 5. Study sponsored by Society of Plastics Industry.

feed. The general combustion performance is shown in Table 7.7. Wet refuse used in test D-1 depressed the arch and flue-gas temperatures. The same effect, but to a lesser extent, was observed in test D-2. Tests D-3 and D-4 proceeded normally; temperatures were high and burnout was good. The addition of PVC increased the odor of the flue gas. There was no unusual clogging of grates during the tests. The concentrations of noxious gases are shown in Table 7.8. A considerable increase in chloride concentration was observed, but no phosgene or free chlorine was detected.

TABLE 7.4 Concentrations of Noxious Chemical Species in the Flue Gases During the Polystyrene Series (ppm by Volume of Dry Gas, Actual and Corrected to 12% CO_2)

Chemical	Test no.		
	B-1 No PS added	B-2 2%	B-3 4%
Cl^-, Actual[a]	223	242	226
Corrected	438	475	450
F^-, Actual	7.0	5.9	6.5
Corrected	13.7	11.6	13.0
CN^-, Actual	<0.01	<0.01	<0.01
Corrected	<0.02	<0.02	<0.01
NO, Actual	35.3	40.2	37.6
Corrected	69.2	79.0	75.0
PO_4^{3-}, Actual	0.54	1.3	0.21
Corrected	1.06	2.6	0.42
SO_2, Actual	44.8	32.4	31.1
Corrected	87.9	63.5	62.0
Organic acids, as CH_3COOH, Actual	41.7	54.1	77.5
Corrected	81.8	106.1	154.5
Aldehydes, as HCHO			
Actual	1.2	1.6	1.4
Corrected	2.4	3.1	2.8

Note: Tests were not run for phosgene and free chlorine, as the polystyrene contained no chlorine.

[a] Average of scrubber and NaOH/impinger methods

Reprinted from Ref. 5. Study sponsored by Society of Plastics Industry.

TABLE 7.5 General Combustion Performance During Polyurethane Series

	Test no.		
	C-1	C-2	C-3
PU added (% refuse)	None	2	4
Refuse (lb/hr)	15,333	13,333	12,000
PU foam	0	267	480
Total (lb/hr)	15,333	13,600	12,480
Furnace arch (°F)	1,431	1,601	1,634
Flue gas (°F)	1,034	1,127	994
Flue gas analysis (Orsat; dry vol %)			
CO_2	5.62	5.57	5.10
CO	0.02	0.03	0.0
O_2	14.65	14.61	15.18
N_2	79.71	79.79	79.72
Burnout from end of grate (ft)	1.3	3.3	6.1
Smoke (Ringelmann)	0.00	0.104	0.15
Odor units of flue gas	3	2	4

Reprinted from Ref. 5. Study sponsored by Society of Plastics Industry.

The study resulted in the following conclusions:

1. Chlorine was found to be present in the normal refuse and in the PU and PVC materials added. During combustion most of the chlorine was evolved as HCl. No free chlorine or phosgene gas was detected. The addition of PE and PS had no effect on the Cl concentration. The addition of PU caused a slight increase and the addition of PVC caused a considerable increase in the concentration of Cl^-.

TABLE 7.6 Concentrations of Noxious Chemical Species in the Flue Gases During the Polyurethane Series (ppm by Volume of Dry Gas, Actual and Corrected to 12% CO_2)

Chemical	Test no.		
	C-1 No PU added	C-2 2%	C-3 4%
Cl^-, Actual[a]	248	320	319
Corrected	530	689	751
F^-, Actual	5.0	16.8	12.6
Corrected	10.7	36.2	29.7
CN^-, Actual	<0.01	<0.01	<0.01
Corrected	<0.02	<0.02	<0.02
NO, Actual	32.2	39.6	39.5
Corrected	68.8	85.3	93.0
PO_4^{3-}, Actual	0.31	4.7	5.5
Corrected	0.66	10.1	12.9
SO_2, Actual	26.1	34.4	32.7
Corrected	55.7	74.0	77.0
Organic acids, as CH_3COOH, Actual	73.9	78.5	73.1
Corrected	158	169	172
Aldehydes, as HCHO			
Actual	5.2	5.1	5.2
Corrected	11.1	11.0	12.2

Note: Tests were not run for phosgene and free chlorine, as the polyurethane contained only a small amount of chlorine.

[a] Average of scrubber and NaOH/impinger methods.

Reprinted from Ref. 5. Study sponsored by Society of Plastics Industry.

2. The average concentration of NO was low in all samples. The addition of plastics caused an increase of NO concentration whether or not the plastic contained nitrogen.
3. There was no correlation between the SO_2 concentration and the amount or type of plastic added.

TABLE 7.7 General Combustion Performance During the Polyvinyl Chloride Series

	Test no.			
	D-1	D-2	D-3	D-4
PVC added (% refuse)	None	2	None	4
Refuse (lb/hr)	15,667	17,000	17,000	14,667
PVC (lb/hr)	0	340	0	587
Total (lb/hr)	15,667	17,340	17,000	15,254
Furnace arch (°F)	1,131	1,585	1,704	1,697
Flue gas (°F)	856	1,024	1,105	1,155
Flue gas analysis (Orsat; dry vol 5)				
CO_2	3.64	4.92	4.99	5.36
CO	0.0	—	0.0	0.0
O_2	16.46	15.13	15.20	14.92
N_2	79.9	79.95	79.81	79.72
Burnout from end of grate (ft)	2	0	12	11.8
Smoke (Ringelmann)	0.189	0.036	0.086	0.043
Odor units of flue gas	20[a]	25	2	20

[a]It should be noted that the burning of normal refuse in this case resulted in an odor level well above that recorded for all the other control samples. The cause was wet refuse that did not produce sufficiently high temperatures to burn all the odorous organic gases.

Reprinted from Ref. 5. Study sponsored by Society of Plastics Industry.

4. There was no correlation between the concentration of CN^- and the amount or type of plastic added.

5. The addition of plastics either had no effect, or caused very slight increases or decreases, in organic acids.

TABLE 7.8 Concentrations of Noxious Chemical Species in the Flue Gases during the Polyvinyl Chloride series (ppm by Volume of Dry Gas, Actual and Corrected to 12% CO_2)

	Test no.			
Chemical	D-1 No PVC added	D-2 2%	D-3 None	D-4 4%
Cl^-, Actual[a]	138	816	304	1354
Corrected	455	1990	732	3030
F^-, Actual	0.92	4.6	2.3	2.8
Corrected	3.03	11.2	5.5	6.3
CN^-, Actual	<0.01	<0.01	<0.01	<0.01
Corrected	<0.03	<0.02	<0.02	<0.02
NO, Actual	25.3	40.6	23.3	46.2
Corrected	83.4	99.1	56.1	103
PO_4^{3-}, Actual	0.27	0.0	0.15	0.0
Corrected	0.89	0.0	0.36	0.0
SO_2, Actual	39.8	37.6	32.9	38.6
Corrected	132	91.7	79.1	86.4
Organic acids as CH_3COOH, Actual	30.3	26.4	48.2	44.4
Corrected	100	64.5	116	99.3
Aldehydes, as HCHO				
Actual	1.8	3.8	1.9	5.2
Corrected	5.9	9.3	4.6	11.6
Phosgene ($COCl_2$), Actual		not detected		
Cl_2		not detected		

[a]Average of scrubber and NaOH/impinger methods.
Reprinted from Ref. 5. Study sponsored by Society of Plastics Industry.

The tests indicate that a conventional incinerator used for the incineration of municipal refuse will perform satisfactorily on normal refuse containing up to 6% or more of plastics [5].

V. INCINERATION OF PREDOMINANTLY PLASTICS WASTE

A. Heat of Combustion

Table 7.9 shows heating values, air requirements, and theoretical flame temperatures of some common plastics. In general, plastics have much higher heating values than municipal refuse (2 to 4 times), require more air for complete combustion, and have higher flame temperatures.

B. Problems Associated with the Incineration of Pure Plastics Waste

Waste plastics can be a valuable source of energy. Although the standard incinerators are capable of handling municipal refuse containing plastics, they cannot usually handle pure plastics waste.

The following problems are associated with the incineration of plastics:

Toxic Gases. When PVC is burnt, HCl gas is generated ($\frac{1}{4}$ to $\frac{1}{2}$ of the weight of the resin), and urethanes generate HCN.

Soot. Imperfect burning of plastics will produce soot. Plastics require 3 to 10 times more combustion air than the average municipal refuse. If they are burned in conventional incinerators, there will be a shortage of oxygen.

Disposal of Ash. Lead and cadmium salts are used as PVC stabilizers. They will remain as ashes containing lead and cadmium, causing disposal problems.

Disposal of Water. HCl is produced by the incineration of PVC. HCl is absorbed in water or by chemicals. Acidified water cannot be disposed of without proper treatment.

Incinerator Damage Caused by Excessive Heat. The temperatures generated during the combustion of plastics are much higher than those generated during the combustion of municipal refuse. The high temperature could cause damage to incinerators.

TABLE 7.9 Heating Values, Air Requirements, and Theoretical Flame Temperature of Some Plastics

Composition of subunit	Subunit MW	Heating value (Btu/lb)	Stoichiometric burning			Common name
			Mol O_2 per mol compound	Ft^3 air/lb 760 torr, 25°C	Theoretical flame temp. (°C/°F)	
C_2H_4	28.05	20,050	3	200	2120/3850	High-density polyethylene
C_2H_4	28.05	20,020	3	200	2120/3850	Low-density polyethylene
C_3H_6	42.08	20,030	4.5	200	2120/3850	Polypropylene, isotactic syndiotactic
C_3H_6	42.08	20,010	4.5	200	2120/3850	Polypropylene, atactic
C_4H_3	56.11	20,060	6	198	2120/3850	Poly-1-butene, isotactic
C_4H_3	56.11	20,150	6	198	2130/3870	Polyisobutylene
	68.12	19,490	7	190	2190/3970	Natural rubber (no sulfur)
C_8H_8	104.2	17,850	10	179	2210/4010	Polystyrene, isotactic
C_8H_8	104.2	17,870	10	179	2210/4010	Polystyrene, atactic—crystal
C_2H_3F	46.05	9,180	2.75	112	1710/3100	Polyvinyl fluoride

Quaternary Recycling

Formula	MW					Name
C_2H_2F	64.04	3,940	2.5	73	1090/2000	Polyvinylidene fluoride
C_2F_4	100.0	144	2	37	0	Polytetrafluoroethylene
C_2F_3Cl	116.5	482	2	32	320/615	Polychlorotrifluoroethylene
C_2H_3Cl	62.50	7,720	2.75	82	1960/3550	Polyvinyl chloride
$C_2H_2Cl_2$	96.95	4,315	2.5	48	1840/3340	Polyvinylidene chloride
C_2H_4O	44.05	11,470	2.5	106	2120/3850	Polyethylene oxide
C_3H_6O	58.08	13,410	4	129	2100/3810	Polypropylene oxide, 27% isotactic
C_3H_6O	58.08	13,400	4	129	2100/3810	Polypropylene oxide, 100% atactic
$C_5H_8O_2$	100.1	11,470	6	112	2070/3760	Polymethyl methacrylate
$C_{16}H_{14}O_3$	254.3	13,310	18	132	2190/3980	Polycarbonate
C_3H_3N	53.06	13,860	3.76	132	1860/3380	Polyacrylonitrile
		7,590 (9)	—	70		Paper
		8,520 (9)	—	80		Woodflour

Incinerator Damage Due to an Insufficient Oxygen Supply. Most of the conventional incinerators are not capable of supplying an adequate amount of air for complete combustion of plastics. During the incomplete combustion, soot is produced which will stick to the pipe walls of the heat exchanging unit, affecting the performance.

Corrosive Damage. Combustion products such as HCl, NH_3, SO_2, SO_3, NO_x, and RCOOH are corrosive and will cause damage to the components of the incinerator. If the waste contains water, it will accelerate the corrosive action of the gases [7].

C. Incinerators Suitable for Plastics Waste

Hishida [7] lists the conditions that must be met if an incinerator is to be used for the combustion of waste plastics.

The incinerator has to be designed for good burning and the prevention of soot delivery.

Walls and furnace beds must be able to withstand the high temperatures generated by the combustion of plastics.

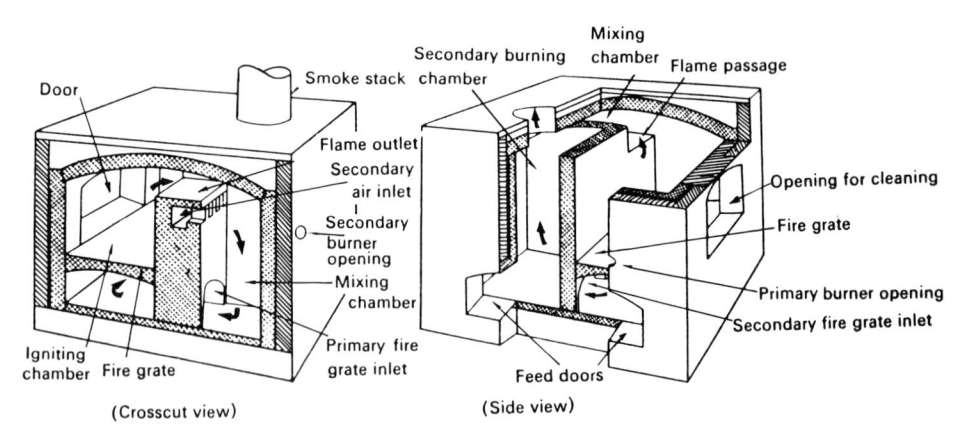

FIG. 7.7 Floor-burning multistage-type incinerator. (Reprinted from Ref. 7, courtesy *Japan Plastics Age*.)

Quaternary Recycling

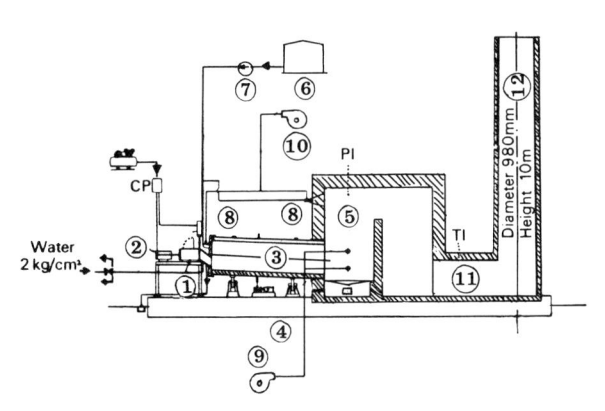

FIG. 7.8 Rotary kiln multistage-type incinerator: (1) feed box; (2) feed drive; (3) rotary kiln; (4) driving unit for rotary kiln; (5) reburning chamber; (6) diesel oil tank; (7) diesel pump; (8) diesel burner; (9) push-in fan; (10) burner fan; (11) smoke passage; (12) smoke stack. (Reprinted from Ref. 7, courtesy *Japan Plastics Age*.)

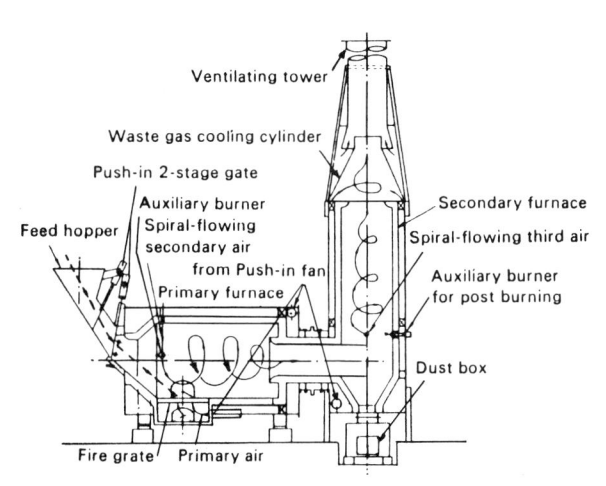

FIG. 7.9 Air-circulating incinerator. (Reprinted from Ref. 7, courtesy *Japan Plastics Age*.)

Plastics wastes

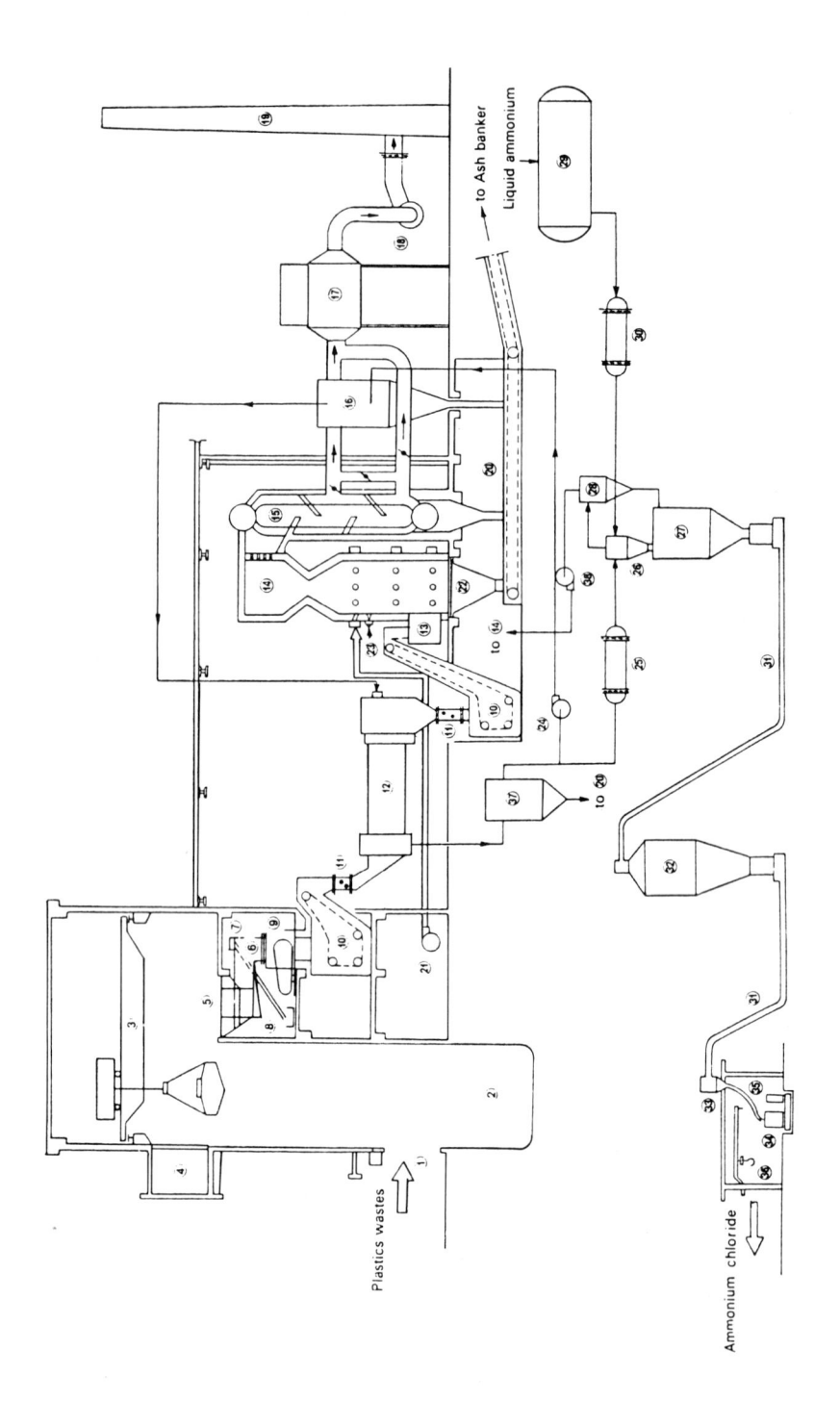

to Ash banker

Liquid ammonium

Ammonium chloride

302

FIG. 7.10 Polymer waste disposal plant: (1) open/close fan; (2) depositioning well; (3) feed crane; (4) crane-operating room; (5) hopper; (6) vibrating feeder; (7) metal collector; (8) conveyor for collected metal; (9) crusher; (10) fixed-quantity feeder; (11) gas seal feeder; (12) pretreatment unit; (13) spreader; (14) incinerator; (15) boiler; (16) circulating gas heater; (17) dust collector; (18) air induction fan; (19) smoke stack; (20); flight conveyor for ash; (21) push-in fan; (22) open/close-type ash-dispensing bed; (23) auxiliary burner; (24) gas-circulating fan; (25) gas cooler; (26) reactor; (27) ammonium chloride separator; (28) gas filter; (29) ammonium storage; (30) ammonium carburetor; (31) conveyor for ammonium chloride; (32) ammonium chloride storage; (33) feed chute; (34) container bag; (35) weight scale; (36) discharge hoist; (37) multicyclone; (38) surplus gas fan. (Reprinted from Ref. 7, courtesy *Japan Plastics Age*.)

Air-supplying equipment has to be able to provide 2.5 to 3 times the amount of air theoretically required for combustion of the plastics.

The incinerator should be designed in such a way as to maintain the temperature below 1150°C.

Since the volume of smoke is proportional to the volume of the air used, the smoke stack must have a larger diameter than that used with a conventional incinerator.

To handle self-extinguishing plastics a preheater must be used.

Feeding equipment capable of handling plastics has to be installed [7].

Examples of incinerators suitable for plastics waste are given below. These can be used for the recovery of energy from plastics, or merely for the disposal of plastics.

Figure 7.7 shows a floor-burning multistage-type incinerator suitable for the batch incineration of plastics. Waste plastics are fed through the feed doors and deposited on the fire grate. The gas ignited on the furnace floor is mixed with primary air. The gases produced during the combustion contain particulate matter as well as incompletely oxidized compounds. Final oxidation takes place in the secondary burning chamber, where the auxiliary burner may be employed if the temperature is low. A continuous rotary kiln-type multistage incinerator developed by Mitsubishi Heavy Industries Ltd. for the incineration of plastics is shown on Fig. 7.8. Primary incineration is accomplished with the kiln slowly rotating. At the start of the operation an auxiliary burner is used to raise the temperature of the furnace. The incineration is completed in the stationary secondary combustion chamber.

Figure 7.9 shows a pressure-type air-circulating incinerator. A horizontal cylinder is used as a primary furnace and a vertical cylinder as a secondary furnace. Increments of waste are continuously fed into the incinerator through the hopper and two-stage gates. In the furnace, in addition to the primary air, secondary air is supplied which circulates along the cylinder, providing a degree of mixing and turbulence. The tertiary air used for the final combustion also flows in a tangential direction.

Figure 7.10 is the schematic of a large incinerator built by the Takuma Boiler Mfg. Co., e specially for the incinceration of PVC. HCl is produced in the rotary kiln by heating PVC waste [12]. After HCl is vaporized from the resin, the carbonized plastic is burned in the incinerator with much less danger of corrosion. Since the volume of gas containing HCl is small, the disposal problems are minimized. HCl gas may be passed through a cyclone followed by a gas cooler, and then reacted with gas to produce ammonium chloride, which can then be collected by a dust collector.

REFERENCES

1. "Municipal Solid Waste. . . A Source of Energy," NCRR Bulletin 3(3), 1973.
2. J. L. Pavoni, J. E. Heer, and D. J. Hagerty, Handbook of Solid Waste Disposal, Van Nostrand Reinhold, New York, 1975, pp. 57-168.
3. C. J. Huang and C. Dalton, Energy Recovery from Solid Waste, Vol. 2, NASA CR 2526, 1974.
4. A. J. Warner, C. H. Parker, and B. Baum, Solid Waste Management of Plastics, Report for Manufacturing Chemists Assoc., De Bell & Richardson, Inc., Washington, 1970.
5. F. R. Kaiser and A. A. Carotti, "Municipal Incineration of Refuse with 2% and 4% Additions of Four Plastics: Polyethylene, Polystyrene, Polyurethane Polyvinyl Chloride," Report to The Society of Plastics Industry, New York, 1971.
6. J. L. Throne and R. G. Griskey, "Heating Values and Thermochemical Properties of Plastics," Mod. Plast. 49(11), 1972.
7. K. Hishida, "Waste Disposal," Japan Plast. Age, 9(11), 1971.

Chapter 8

DISPOSAL OF WASTE PLASTICS WITHOUT THE RECOVERY OF VALUE

I. INCINERATION WITHOUT THE RECOVERY OF ENERGY

The most widely used method of managing municipal refuse is straight disposal without recovery of value. The methods used are incineration without recovery of energy and landfill.

Incineration of solid municipal refuse can be performed in an incinerator or by open burning. Open burning on level ground or in pits was one of the earliest methods of disposals, but because of severe air pollution problems this method is rarely used today. The combustion of municipal refuse and plastics waste in incinerators is described in Chap. 7.

II. LANDFILL

Landfill is the most common method of municipal solid waste disposal. It is also the only "final" method, landfill being the ultimate destination of the waste. In the process of landfilling, a number of potentially valuable components are lost. There are two main techniques for landfilling: open dumping and sanitary landfill.

A. Open Dumping

Open dumping was the original method of waste disposal. In this method refuse is deposited in an open area and allowed to decompose.

Its sole advantage, a relatively low disposal cost, is far outweighed by the disadvantages: odor, the scattering of lightweight wastes by the wind, and the presence of pests such as rats and mice. Open dumping is still being practiced, but its importance is diminishing.

2. Sanitary Landfill

Sanitary landfill is a modern method of landfilling. Warner et al. [1] quote the definition proposed by the American Society of Civil Engineers: "Sanitary landfill is a method of disposing of refuse on land without creating nuisances or hazards to public health or safety, by utilizing the principles of engineering to confine the refuse to the smallest practical area, to reduce it to the smallest practical volume and to cover it with a layer of earth at the conclusion of each day's operation or at such more frequent intervals as may be necessary."

The process of sanitary landfill may be broken down into four basic operations:

1. The solid wastes are deposited in a controlled manner in a prepared area of the site.
2. The solid wastes are spread and compacted in thin layers.
3. The solid wastes are covered daily, or more frequently if necessary, with a layer of earth.
4. The cover material is compacted daily.

Ultimately, a sanitary landfill site may be used for recreational, agricultural, or commercial use. Recreational use is so far the most popular. Solid wastes are spread in thin layers approximately 2 ft thick, and then compacted. The degree of compaction depends on the type of solid waste, the weight and type of compacting equipment, and the number of passes of compacting equipment. Usually a layer of compacted sandy loam or other soil about 6 in. in depth is deposited to cover a daily deposit of the refuse. A minimum of 24 in. of compacted soil is recommended as a final cover. In order to prevent water from eroding the cover or seeping into the landfill, good drainage must be provided. Depending on the type of refuse placement, sanitary landfill procedures may be classified into trench, area, and ramp methods. In the trench method a long and narrow excavation is made and the removed soil stockpiled. Wastes are deposited at one end of the trench, spread at a small inclination, and compacted. At the end of the day part of the stockpiled soil is used to cover the refuse. When the entire trench has been filled, a final layer of soil is placed over the top of the completed deposit.

In the area method the site does not require any preparation; however, the top soil may be removed if desired. The refuse is dumped on the ground, spread, and compacted. The refuse is covered with a layer of soil at the end of each day and with a thicker layer after the deposition area has been filled. The cover soil is normally transported from another location.

The ramp method combines some features of the two methods just described. Initially a small excavation is made. Refuse is deposited on the face of the slope, spread, compacted, and then covered with soil. The process is repeated at the face of each newly created slope. Two or three different landfill methods may be used on the same disposal site.

Sanitary landfill is a low-cost method of disposal, but under certain conditions the burial of solid waste has a potential for chemical and bacteriological pollution of ground and surface water. Such water pollution can be prevented by: (1) Locating disposal areas away from water sources such as wells, lakes, and streams; (2) avoiding landfilling above the subsurface as it could lead to the transfer of leachate from landfill to water resources; (3) using earth cover that is relatively impervious to water; and (4) providing good surface drainage.

Sanitary landfills produce methane and other gases, and must be designed so as to enable the escape of the gas to atmosphere. If the escape of gas is prevented, it might be forced into sewers serving homes in the area, thus causing explosions. Landfill fires also present control problems. The potential for fire can be minimized by proper dumping procedures and by leaving the minimum amount of area open at any time [1,2].

III. PLASTICS IN LANDFILL

There are two opposing views on the role of plastics refuse in landfill: Plastics reduce the effectiveness of the disposal process. Plastics waste is much more bulky than other materials, and on the same weight basis require considerably more disposal space. They also decompose very slowly and thus increase the time necessary for reuse of the landfill site.

On the other hand, plastics do not reduce the effectiveness of the disposal process. Under the weight of the compacting equipment plastics are compacted to the same density as other materials. Plastics are not soluble in water and decompose slowly; their decomposition products are usually inert and thus there is no polluting of water resources.

Unfortunately, there are very few data available on the behavior of plastics in landfill, so that it is impossible to confirm or refute either of these views.

Plastics in sanitary landfill decompose because of the combined action of such mechanisms as oxidation, hydrolysis, attack by microorganisms, stress cracking, and attacking by insects and rodents [3].

Simple oxidation is probably most significant to polyolefins. Elevated temperature due to the fermentation of refuse increases the rate of the oxidation reaction. Conditions existing in landfill might also cause a leaching out of antioxidants from the plastics, and thus accelerate the oxidation process.

Hydrolysis affects such plastics as polyesters, polyurethanes, and cellulose derivatives. As with oxidation, the elevated temperature of the landfill increases hydrolysis reaction rates.

Biodegradation is the attack and assimilation of the material by microorganisms such as fungi and bacteria. Potts [4] investigated the susceptibility of various polymers to biodegradation by placing test specimens in solid agar growth media deficient only in carbon. Any growth that would occur under these conditions is dependent on the utilization by the organism of some component of the specimen as a carbon source. The media and specimens were inoculated with the test microorganisms and incubated for 3 weeks. After various exposure times (up to the 3 weeks) samples were examined and assigned the following growth ratings:

0 — no growth
1 — traces (less than 10% covered)
2 — light growth (10 to 30% covered)
3 — medium growth (30 to 60% covered)
4 — heavy growth (60 to 100% covered

Table 8.1 shows the results of tests carried out on some of the common plastics additives. It is important to note that some PVC plasticizers are biodegradable and their presence can be expected to increase the rate of degradation of PVC. Table 8.2 shows the results of tests conducted on commercial plastics. Nearly all the plastics tested were resistant to biodegradation. Several samples which did show susceptibility to attack provided no growth after the sample was extracted with solvent. This behavior indicates it is the additive which is attacked by microorganisms and not the polymer itself. The study conducted on straight-chain hydrocarbons of various molecular weights demonstrated that in order to be attacked by microorganisms the molecular weight has to be below about 500. Table 8.3 shows the effect of the average molecular weight of HDPE and LDPE on biodegradability. It is assumed that

TABLE 8.1 Biodegradability of Plastics Additives

Trade name	Chemical name or type	Growth rating
Antioxidants		
BHT	Hindered phenol	0
Santonox R	Hindered phenol	0
Topanol CA	Hindered phenol	0
Irganox 1010	Hindered phenol	0
Polygard	Nonyl phenyl phosphite	Z.I.[a]
Slip agents		
Erucamide	C_{22} primary amide	4
Stearamide	C_{18} primary amide	4
HTSA-1	Olealyl palmitamide	2
UV absorbers		
Eastman DOBP	2-OH-4 dodecyloxy benzophenone	0
Eastman OPS	p-Octylphanyl salicylate	0
Plasticizers		
Flexol DOP	Di-2-ethylhexyl phthalate	0
Flexol TCP	Tricresyl phosphate	1
Flexol EPO	Epoxidized soybean oil	4
Plastolein 9765	Aliphatic polyester	4

[a] Zone of inhibition—fungicidal.
Reprinted from Ref. 4, courtesy Institute of Electrical and Electronics Engineers.

only the low molecular weight fraction of the polymer is being attacked. Similar tests conducted on PS showed no attack. Similar results were obtained on some of the ethylene and styrene copolymers. In general, plastics show a relatively good resistance to microbiological attack [4]. This resistance may be reduced by the incorporation of biodegradable fillers (such as starch) or some plasticizers.

TABLE 8.2 Biodegradability of Commercial Plastics

Product	Growth rating
Polyethylene household wrap	2
The above extracted with toluene	1
PVC—epoxidized soybean oil plasticizer	3
The above extracted with toluene	1
Polypropylene	1
Polystyrene	1
Polyethylene terephthalate (Mylar)	1
Polyvinylidene chloride	1
Acrylonitrile-butadiene-styrene copolymer (ABS)	0
ABS-polycarbonate blend	0
Butadiene-acrylonitrile rubber	0
Styrene-acrylonitrile copolymer	0
Rubber modified polystyrene	0
Styrene-butadiene block copolymer	1
Polymethyl methacrylate	0
Rubber-modified polymethyl methacrylate	0
Polyethylene terephthalate	0
Polycyclohexanedimethanol terephthalate	0
Bisphenol A polycarbonate	0
Bisphenol A polysulfone	0
Poly(4-methyl-1-pentene)	0
Polyisobutylene	0
Chlorosulfonated polyethylene	0
Cellulose acetate or butyrate	0
Nylon-6, nylon-6/6, nylon-12	0
Polyurethane	4
Polyvinyl butyral	0
Polyformaldehyde	0
Polyvinyl ethyl ether	0
Polyvinyl acetate	0

Reprinted from Ref. 4, courtesy Institute of Electrical and Electronics Engineers.

TABLE 8.3 Effect of PE Molecular Weight on Biodegradability

Product type	Mol wt	Growth rating
HDPE	10,970	2
HDPE	13,800	2
HDPE	31,600	0
HDPE	52,500	0
HDPE	97,300	1
LDPE	1,350	1
LDPE	2,600	3
LDPE	12,000	2
LDPE	21,000	1
LDPE	28,000	0

Reprinted from Ref. 4, courtesy Institute of Electrical and Electronics Engineers.

Stress cracking is the cracking of plastics subjected to certain chemical environments while under a physical stress. It can be assumed that situations conducive to stress cracking exist in landfill, and that they cause breakdown of at least some plastics into smaller pieces.

Direct animal attack is difficult to quantify, but because of the large number of animals attracted to landfill sites it is assumed to be significant [3].

REFERENCES

1. A. J. Warner, C. H, Parker, and B. Baum, *Solid Waste Management of Plastics*, Report for Manufacturing Chemists Association, DeBell Richardson, Inc., Washington, 1970.
2. J. L. Pavoni, J. E. Heer, and D. J. Hagerty, *Handbook of Solid Waste Disposal*, Van Nostrand Reinhold, New York 1975, pp. 169-223.
3. G. J. L. Griffin, "Biodegradable Fillers in Thermoplastics," *Conf. on Degradability of Polymers and Plastics,* IEE, London, 1973.
4. J. E. Potts, "The Effect of Chemical Structure on the Biodegradability of Plastics," *Conf. on Degradability of Polymers and Plastics*, IEE, London, 1973.

INDEX

ABS,43,122,209
Acrylics,44
Aminoplastics,16
Antioxidants,2,20
Antistatic agents,2
Assembler,65,67
Automatic Scrap Recycle System,152

Biodegradation,310
Black-Clawson process,92, 100,109
Blends,158,159
Blow molding,33
Broker,56

Cable extrusion,32
Calendering,36,68
Cellulosics,45
Chain fragmentation,235
Chippers,85
Chlorinated polyethylene, 186-188
Coextrusion,30,206
Coinjection,206,207,209
Colorants,19
Compatibilizers,186
Compounder,56,65,66,70
Compounding,28
Compression molding,36
Copolymer,3
Condux Plastcompactor,147
Converter,67,70,179
Crushers,85

Decomposition,264,273
Degradation,119,121
Depolymerization,235,274,275
Disk mills,85,88
Disposal area,79
Distributor,67,73
Drum pulverizers,85,86

Engineering resins,2,14
Epoxy,17,44,215
Ethylene vinyl acetate,45
Extruder,27,147
Extrusion,27,68,204

Fabricator,56,65,66,67,70,73
Fillers,23-25
Film extrusion,30
Fläkt waste recovery system,94
Flame retardants,21
Flita system,177
Flow modifiers,18
FN machine,176,177
Furnace,281,282

Glass fibers,2
Glass transition,6
Glycolysis,271
Granulation,135
Granulators,134,214
Grinding,143,145

Hammermills,85,89,214
Homogenization,159
Homopolymers,3

315

Index

Hydrolysis, 265, 266, 270
Hull Corporation, 214

Impact modifiers, 18
Incineration, 67, 80, 82, 219, 278, 279, 297, 304, 306
Incinerator, 279, 281, 285, 286, 288, 300, 304, 305
Industrial plastics waste, 64, 67, 68
Injection molding, 32, 68
Input analysis, 73

Japan Steel Works Ltd., 183

Kabor Ltd., 183
K Board, 183
Klobbie, 174, 175

Lamination, 39
Landfill, 67, 81, 306-309
Lubricants, 19

Manufacturer, 56, 73
Mitsubishi, 169
Molecular weight, 3
Municipal refuse, 79, 278, 306

Nikko Waste Plastics Reclamation line, 183
Nuisance plastics, 64, 65, 67

Open burning, 80
Open dump, 80
Open dumping, 306
Output analysis, 73, 77

Packager, 65, 67, 73
Packaging, 44
Pallman Plast-Agglomerator, 147
Pelletizing, 28
Phenolics, 15, 209, 213, 214
Pigments, 2
Pipe, 29
Plasticizers, 19, 20
Plastics waste, 70, 73, 145, 156, 157, 159, 186, 203, 235, 297, 300, 304
Polybutylene, 49
Polyester, 16, 54, 215, 265, 274, 275, 310

Polyethylene, 1, 2, 7, 14, 15, 47, 59, 111, 122, 159, 161, 170, 186, 188-190, 192, 205, 238, 239, 241, 249, 250, 254, 257, 259, 288, 293
Polyethylene terephtalate, 49, 274
Polymethyl methacrylate, 2, 235, 254
Polyolefins, 108, 117 (see also Polyethylene, Polypropylene)
Polypropylene, 2, 10, 48, 50, 111, 122, 249, 254
Polystyrene, 1, 12-14, 47, 48, 51, 59, 108, 111, 117, 122, 161, 170, 186, 188, 188-190, 239, 241, 254, 257, 259, 264, 288, 293, 311
Polysulfones, 54
Polyurethane, 17, 50, 51, 170, 264, 265, 267, 270-273, 288, 289, 293, 310
Polyvinyl chloride, 1, 10, 43, 49, 50, 51, 59, 103, 105, 108, 110, 111, 117, 161, 170, 186, 239, 241, 243, 257, 259, 288, 290, 291, 293, 297, 310
Postconsumer plastics waste, 64 (see also Plastics waste)
Prices of plastics, 57-59, 61
Process Control Corporation, 152
Profiles, 29
Pulpers, 85, 88
Pulverizers, 214
Pyrolysis, 82, 218, 219, 221, 224, 228, 231, 234-237, 247, 250, 254, 255, 257, 258, 264
Pyrolysis reactor, 226, 228, 231, 234-236, 255

Rasp mills, 85, 86
Reaction injection molding, 50
Reclamat International Ltd., 182
Recycler, 214, 215
Recycling, 119, 151, 156, 158, 168
Regal Packaging, 179
Reinforcements, 26
Release agents, 19
Remaker, 177
Reprocessing, 157
Reprocessor, 66, 67, 70
Resin manufacturer, 65, 66

Reverzer,169,172
Rotational molding,36,68

Sanitary landfill,80
Scrap plastics,64,65,67,73
Separation,85,89,91,92,97,
  100,101,103,106-108,110,
  112,157
Shears,85,86
Sheet extrusion,29
Size reduction,85
Solid waste,78,108
Specialty thermoplastics,14

Thermoforming,38
Thermoplastics,5,6
Thermosets,6,15
Transfer molding,36
Tubing,29
Tupbord,182

Ultraviolet stabilizers,2,21
Union Carbide plastics
  pyrolysis system,247

Weist Industries, Inc.,150
Wire extrusion,32

UCC South Charleston, WV 770
TD798 .L23
Leidner, Jacob   /Plastics waste

1015044

TABLE 1.14 Plastics in Furniture, 1978

Material	Consumption (1000 Tons)
ABS	7
Melamine	8
Phenolic	
Decorative laminates	10
Plywood	20
Polyester	7
PE	8
PP	14
PS	46
Polyurethane	
Flexible foam	227
Rigid foam	13
Other	7
PVC	105
Other	4
Total	476

Reprinted from Ref. 3, courtesy McGraw-Hill.

E. Electrical and Electronics Uses

Plastics have been used for some time by the electrical and electronics industries in such applications as wire and cable insulation, circuit boards, and housings. The present expansion of this market is caused mainly by the increasing requirements for plastics in the data-processing segment of the industry. Table 1.15 summarizes uses of plastics by the electrical and electronics industries.